# Religion on Our Campuses

## A Professor's Guide to Communities, Conflicts, and Promising Conversations

*Mark U. Edwards, Jr.*

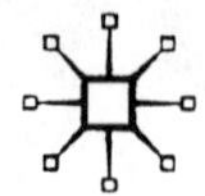

RELIGION ON OUR CAMPUSES



First published in 2006 by
PALGRAVE MACMILLAN™
175 Fifth Avenue, New York, N.Y. 10010 and
Houndmills, Basingstoke, Hampshire, England RG21 6XS
Companies and representatives throughout the world.

PALGRAVE MACMILLAN is the global academic imprint of the Palgrave Macmillan division of St. Martin's Press, LLC and of Palgrave Macmillan Ltd. Macmillan® is a registered trademark in the United States, United Kingdom and other countries. Palgrave is a registered trademark in the European Union and other countries.

ISBN-13: 978–1–4039–7209–5 hardback
ISBN-10: 1–4039–7209–5 hardback
ISBN-13: 978–1–4039–7210–1 paperback
ISBN-10: 1–4039–7210–9 paperback

Library of Congress Cataloging-in-Publication Data
Edwards, Mark U.
Religion on our campuses : a professor's guide to the communities, conflicts, and promising conversations / Mark U. Edwards, Jr.
p. cm.
Includes index.
ISBN 1–4039–7209–5—ISBN 1–4039–7210–9
1. Universities and colleges—United States—Religion. 2. College teachers—Religious life—United States. I. Title.
BV1610.E39 2006
200.71′173—dc22 2005056676

A catalogue record for this book is available from the British Library.

Design by Newgen Imaging Systems (P) Ltd., Chennai, India.

First edition: September 2006

10 9 8 7 6 5 4 3 2 1

Printed in the United States of America.

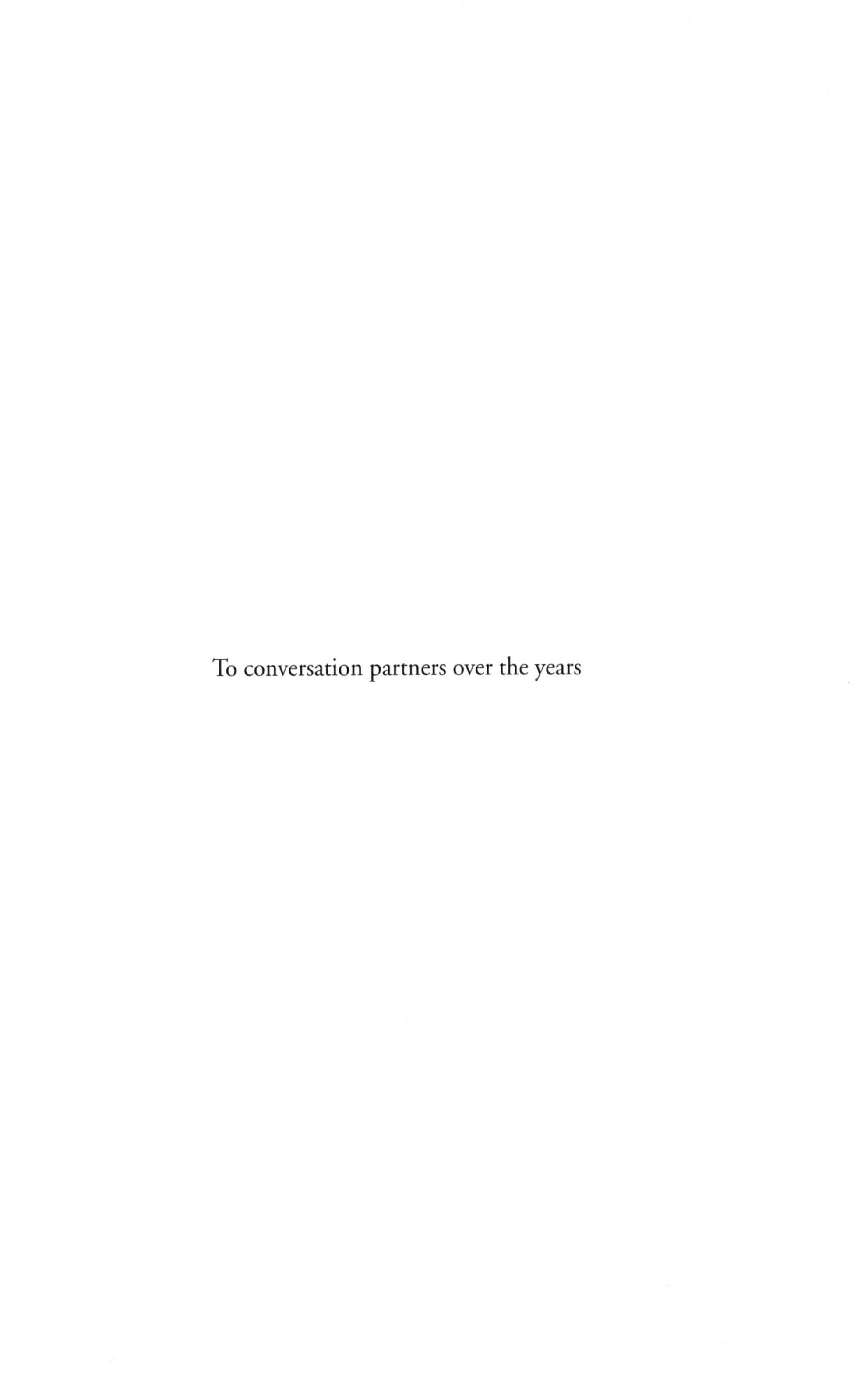

To conversation partners over the years

# Contents

# Preface

With this book I seek to help faculty and administrators better understand attitudes toward religious discourse in scholarship and teaching. My angle of entry is to stress the formative (and "conformative") role of professional disciplinary communities in establishing standards and practices for professional discourse regarding religion. My commended approach is to deal with these issues conversationally. My ultimate goal is to open up space for a circumspect readmission of religious discourse into scholarship and teaching.

I offer a model for understanding disciplinary formation and contrast that model with (nowadays, generally much weaker) formation into a religious tradition. Along the way I sketch the genesis of professional disciplinary communities and their relationship to the religious beliefs and practices of their day. I suggest ways that faculty can explore their own disciplinary formation.

Disciplinary communities set and enforce standards for scholarship and teaching. This collective enterprise advances our understanding of self and world and secures our authority and freedom as scholars and teachers. But along with these positive goods come limits and responsibilities that have tended, for reasons both good and bad, to exclude explicitly religious discourse from most scholarship and much teaching. This is especially true in secular colleges and universities, which are my primary focus. Against this backdrop, I suggest how circumspect religious discourse might appropriately reenter the conversation in ways consistent with the ongoing arguments about standards and practices that distinguish all living disciplinary traditions.

In discussing appropriate religious discourse on campus, I distinguish between biographical disclosure of religious conviction and explicitly religious claims or warrants employed to justify a scholarly position. Biographical disclosure, when deftly handled, can help colleagues and students understand where particular faculty members are "coming from" and what motivates them as professional disciplinary scholars. I see a broad, but necessarily cautious role for religious self-disclosure in advancing understanding within the academy. I see a far more limited role for explicitly religious claims or warrants. If advanced at all, they are best limited to moral claims,

claims regarding human nature, and claims regarding maximally comprehensive views of reality (i.e., to certain metaphysical claims). In deference to the standards and practices of most academic discourse, self-disclosure or explicitly religious claims should be limited to cases where the religious perspective actually furthers the conversation and contributes to deepened understanding. I offer suggestions for evaluating when self-disclosure or explicitly religious claims may be appropriate, and when not. Throughout, I hold up the risk as well as the benefit of allowing religious discourse back into the academy.

My focus is on faculty—on faculty formation, faculty self-understanding, faculty conversation with other faculty—and on the standards that our disciplinary communities instill in us as faculty and what these standards may mean for religious perspectives in scholarship and teaching. I deal with students in relation to the work of faculty as disciplinary professionals and teachers. Those interested in student religiousness per se should consult the books and surveys I cite along the way.

I tried out precursors of these chapters in several faculty seminars around the country. On the basis of that experience, I am convinced that to best understand and appreciate the formative influence we have undergone as disciplinary professionals, we need to compare experiences and perspectives with colleagues in other disciplines and from different religious or secular traditions. This contention is central to my project and informs every chapter. Those willing to converse civilly and openly with colleagues about these issues are best positioned to understand the real issues and to decide collegially what the appropriate role of explicitly religious discourse should be on campus.

In this book I am commending a conversational approach to the difficult and conflicted issue of religious perspectives in scholarship and teaching. I rely on conversations to provide insight and advance my argument. These conversations may not only be internal, between reader and author, but they may also proceed face to face, among faculty from different disciplines assembled to learn from each other. Conversation lies at the heart of this project, in its genesis, in its recommended approach, and in its hoped-for outcome.

Over the course of my career, I have been formed intellectually by crucial conversations with colleagues. In these collegial exchanges, I came to see the power that distinguishes conversation from argument or debate. Over the course of writing this book, I have benefited from many face-to-face conversations with colleagues at several institutions. I recognize that I have captured but a fraction of the experience, insight, and wisdom that others have shared with me. So I need both to thank my conversation partners for their generous sharing of their own perspectives regarding religious perspectives in scholarship and teaching and to apologize to them for my shortcomings in grasping all they had to say.

Several communities introduced me to the joys and insight that arise from collegial conversation among the disciplines. In the early 1970s, the University of Michigan Society of Fellows gave me an opportunity to converse at length with a group of engaged, open-minded young scholars who were working at the leading edge of their disciplines. In the mid- to late 1970s, colleagues in the Wellesley College faculty colloquium included me in an ongoing conversation about disciplinary distinctiveness and its pedagogical challenges. At my next place of employment, Purdue University, conversations with my colleagues in the history department helped confirm my sense that institutional context shapes faculty and disciplines in powerful ways. When I moved to Harvard Divinity School, I entered into ongoing scholarly conversations about the world's great religious traditions and their interaction. My stint as president of St. Olaf College helped me see that conversations across disciplines must be actively fostered. They don't just happen. The busyness of our work as faculty easily obstructs conversational exchanges across disciplinary divides on issues more intellectually engaging than disputes over general education requirements or promotion decisions. Conversations in these communities have formed me as an academic and a scholar. They inform the writing of this book.

Several colleagues acted as advisors and conversation partners in this project. I especially want to thank four colleagues who have patiently worked with me from the beginning as I struggled toward an appropriate formulation for my concerns: Edward Farley, Mark Schwehn, Ronald Thiemann, and Alan Wolfe. They helped me find my own vision, even as they themselves may have seen things differently. Several readers at various stages in the genesis of the project also enriched and furthered my internal conversation. I especially want to thank Mary Jo Bane, Brent Coffin, Kimberlee Maphis Early, James L. Heft, Richard J. Mouw, Margaret R. Miles, Anne E. Monius, William E. Paden, Richard Parker, Stephanie Paulsell, James L. Pence, William C. Placher, H. Paul Santmire, Douglas Stone, and Julie Boatright Wilson. My daughter, Teon Elizabeth Edwards, kindly helped me see some issues from another generation's perspective.

Martin E. Marty has been a conversation partner extraordinaire. For a time we together—I as president and he as chair of the Board of Regents—conversed regularly about how to further the mission, financial health, and engaged faculty conversations that we thought should distinguish St. Olaf College. Subsequently, we have continued to converse about the manifold character of religious perspectives in the work of the academy. Marty's insights inform many of the arguments I advance in this book, especially the appropriate role for conversation when dealing with conflicted issues of central importance. His influence on me has been profound, but he is certainly not responsible for any errors I've committed in appropriating and deploying his insights.

Three campuses used an earlier version of this book to set the table for a faculty seminar. Listening to colleagues at each of these institutions tell their stories helped me shape the final version. I want to thank the participating faculty at the College of Wooster, Hendrix College, and Willamette University, and especially the seminar conveners Dianna R. Kardulias, Peg Falls-Corbitt, and Karen L. Wood. I also want to thank faculty and administrators at Calvin College, Furman University, Massachusetts Institute of Technology, St. Olaf College, and Wabash College who provided me with crucial early insight into the promise and perils of religious perspectives in scholarship and teaching in today's colleges and universities.

Finally, I want to thank the Lilly Endowment for its generosity, understanding, and support for this project. I particularly appreciate the sound advice and friendly encouragement I have received from Craig Dykstra and Chris Coble. At crucial points along the way they have made all the difference.

# Introduction

As faculty we recognize that religion plays a crucial role in national and world politics, economics, and social relations. We hardly need to be convinced that given the increasing presence in America of adherents to all the great world religions, a well-educated citizen needs to understand religious variety even if she never leaves the United States. We realize that from abortion to civil rights, religious conviction impels American citizens to organize, protest, and engage in often fiercely partisan politics. We hear daily about major religious traditions—Christian, Islamic, Jewish, Hindu, Buddhist—clashing with each other or with modern culture and the emerging global economic system. We know that world literatures, philosophies, and languages reflect religious conviction and offer views of the human, the world, and the divine that give meaning, purpose, and value to life. As a result, in most of our colleges and universities, religious issues and traditions are regularly addressed in courses from political science and sociology to literature and philosophy. In addition, many colleges and universities have a Religious Studies Department that is dedicated to the academic study of religious traditions.

We also recognize that religious beliefs and practices can underlie core elements of personal identity. When challenged, such beliefs can lead to explosions and may underwrite coercion, discrimination, and even violence.

But even as we recognize the importance of religion and its explosive potential, in the academy, we tend to treat religion as something "out there," something to be studied. It is not something "in here," in the academy, that needs to be taken seriously because it shapes how some faculty and students understand the world. We concede its importance out in the world; we see the risks it poses. But we're reluctant to admit its importance within the academy, perhaps because we see only too well its perilous potential. This reluctance has a powerful rationale.

## Conflict

For starters, religion has been demonstrably dangerous to the liberal ideals of higher education. From the late nineteenth century to well into the latter half of the twentieth century, scholars have found themselves grappling with a certain form of Protestant Christianity that inhibited intellectual progress, threatened academic freedom, encouraged discrimination against religious minorities (especially Jews and Catholics), and promoted seemingly insoluble divisions on moral matters.[1] Restraint regarding personal conviction may seem an appropriate price to pay to assure that such things do not recur.

There is also concern that the mention of religious or spiritual convictions may indulge in an undue subjectivity. "Aren't we violating the canons of good, objective scholarship," the conscientious scholar worries, "if we offer religious grounds for accepting a scholarly claim—or perhaps even if we mention our religious convictions in passing?" Don't science and reason offer more compelling and universally accessible alternatives?

Finally, there is the question of good taste. In many academic circles, the mention of religious conviction is a conversation stopper. "So why bring that up?" comes the retort. "We were discussing X or Y, not your private life. Don't bother us with matters that are not our concern."[2]

For these and other reasons (some compelling, some less so), there is a widespread reluctance to mention personal religious conviction in academic discourse. Arguably, this is, on balance, a good thing. But perhaps only on balance. There are losses as well as gains entailed in this restraint.

Not mentioning religious (or analogous) convictions does not make them go away. They are still present. They still work their influence, but perhaps without appropriate examination, discussion, and compensating adjustment. In other words, in exercising such restraint, there may well be a loss of critical self-awareness.

Such a restraint may also be doing our students a pedagogical disservice. If we fail to discuss or even mention the role that deep personal convictions may play in career choices and scholarly interpretations, we may be tacitly encouraging our students to conclude that they don't have to worry about such things. We then forego a splendid opportunity to illustrate the hard work that scholars undertake to identify subjective inclinations and compensate for improper effects.

Sometimes, and for some faculty, the best reason we can give for what we believe on a particular issue and why we believe it is frankly religious. This is especially true for moral judgments within our discipline and disciplinary claims regarding human nature and maximal understandings of reality (i.e., certain metaphysical claims). In such cases, why not simply say so? To pretend otherwise can be disingenuous and unhelpful to the task of advancing understanding and knowledge.

Finally, by acquiescing in a ban on the discussion of religious influences in higher education, we lose an opportunity to understand and respond to what other colleagues really think and why they think it. We also lose a potentially productive opportunity to get to know each other and each other's disciplines better. Consider how seldom we faculty get the chance to engage colleagues across the range of disciplines on matters of intellectual and disciplinary substance where each scholar and each discipline has as much to contribute as the other. We may discuss each other's disciplinary standards in school-wide tenure and promotion committees. We may debate disciplinary approaches when crafting general education requirements. In what other circumstances would a strong mix of the personal and the professional arise?

In the end the deciding question must be, "Is the recognized risk worth the potential gain?"

## A Conversational Approach

It is the contention of this book that a conversational approach offers the greatest potential gain at the least potential risk. A conversational approach may mean, literally, that faculty approach these issues through conversations with colleagues.[3] But a conversational approach can also mean that individual faculty members engage the issues primarily in order to deepen their own understanding. Conversation as advocated here sets the ground rules for engaging with a topic that *is* risky and dangerous with the goal of minimizing that risk (at the same time recognizing that risk and danger remain ever present just under the surface of even the most respectful conversations on this explosive subject).

First, the goals of both conversation and a conversational way of knowing is deepened understanding, not agreement or resolution. This is conversation in a sense analogous to the one intended by the philosopher Michael Oakeshott.[4] Participants in a conversation, Oakeshott suggested, are engaged in an activity that may include inquiry, argument, and debate but that ultimately aims at something else. In a conversation, "there is no 'truth' to be discovered, no proposition to be proved, no conclusion sought." Participants "are not concerned to inform, to persuade, or to refute one another, and therefore the cogency of their utterances does not depend upon their all speaking in the same idiom; they may differ without disagreeing.

Of course, a conversation may have passages of argument and a speaker is not forbidden to be demonstrative; but reasoning is neither sovereign nor alone, and the conversation itself does not compose an argument."[5] Arguments may arise in conversations, but the goal of a conversation is not to win (or avoid losing) an argument, but rather to understand better (and perhaps even empathize with) another's position even as one explains (and perhaps even hopes for empathy regarding) one's own position.

Oakeshott's distinction between a conversation and an argument is important. As historian Martin Marty points out, in arguments, the contenders claim to know the answers. They debate with an intent to convince or overthrow those of other opinion. Whereas in a conversation, the interlocutors have questions. They converse with an intent of deepening empathy and broadening understanding.[6] Again in Oakeshott's words, conversation is "not a contest where a winner gets a prize." Rather it is "an unrehearsed intellectual adventure."[7]

Conversations come with a set of assumptions or expectations that distinguish them from debates or arguments. For example, in conversations, everyone is considered more or less equal in expertise and authority. Each person in a conversation has the right to his or her say. No one is an expert whose authority trumps all others. Each has a right to call for reflection, to pose questions, to try to steer the conversation in any direction. In a conversation there is little hierarchy and no arbiter of who's right and who's wrong. Even in the "lightly structured and regulated" optional conversations advocated in the appendix to this book, the leader seeks primarily to keep conversation conversation. The crucial thing is to avoid the academic inclination to turn matters of disagreement into a contest to establish one's own position and tear down another's.

Conversations are situated. They involve specific individuals in distinct contexts seeking to understand each other. Not only do conversations engage contingent, situated conversationalists, but they also frequently deal with contingent, situated people, events, and things. The situated, contingent nature of a conversation—both in who converses and what they converse about—often takes the form of telling stories or of relating anecdotes. We offer the reasons that incline us or others this way or that, and explain why. We tell what was intended, what actually happened, and why. We relate specific events and see meaning (or its lack) in them. Fragments of contingent narrative are the common stuff of our everyday conversations, whether at work or at play, in the academy or at home, in the privacy of our own minds or in public chitchat.

In conversations it is appropriate to bring up feelings as well as ideas, to share that which is subjective as well as objective—assuming that such a distinction can easily be drawn. Since the goal is to deepen understanding of the other (as well as of one's self), expressions of passion, aversion, or indifference have as much right in conversations as claims of fact or narratives of experience. We may attempt to bracket our feelings when doing our scholarship or teaching our students, but we need not exercise such restraint when conversing about what convicts and convinces us—or what convicts and convinces others. In fact, if we fail to include the emotional with the notional, we are likely to shortchange the understanding and social engagement that good conversation aims at.

Conversations are richest (but also, perhaps, most scary) when a diversity of perspectives is present. For his part, Oakeshott insists that conversation is, properly speaking, "impossible in the absence of a diversity of voices: in it different universes of discourse meet, acknowledge each other and enjoy an oblique relationship which neither requires nor forecasts their being assimilated to one another."[8] If Oakeshott is right about this—and I think he is—then conversations on religious commitment within the academy should include faculty who come from different disciplines and who bring varied perspectives and experiences regarding religious belief and practice and the life of the mind.

Conversation in the sense being advocated here is a contingent, emergent way of knowing that is dependent more upon a cooperative interchange than upon a universal logic or the "truth of matters," whatever that might be. The "logic" and direction of a conversation arise out of the skillful use of concrete practices such as story telling, taking turns in conversing, reflecting on what one has heard, posing clarifying questions, and offering analogs and contrasting examples. It is a dialectical game in which all parties are constantly adjusting to the give and take of the conversation.

To make a conversation a dialectical game, certain low-level practices are employed, for example, turn taking. In a true conversation, each person gets his or her turn to speak while others listen. And in good conversational practice, turn taking is more than a serial monolog. Dynamic interaction distinguishes each exchange. Conversationalists regularly feed back what they are hearing (or think they are hearing) from their conversational partners, such as, "You must have felt good about that!"; "Boy, I would have been annoyed if my mentor had done that to me."; "I'm also worried about Susan's attitude . . ." They ask clarifying questions: "Do you really see the choice that starkly?"; "What about Bill's opinion on that?"; "I don't understand what you mean by . . ." They juxtapose experiences: "I had a similar experience with Juan . . ."; "I never thought of it that way, but . . ."; "It wasn't like that at my university . . ." At the very least, they punctuate their partner's account with phatic noises that show that they're following the conservation and that urges their partner on: "Yes, uh hum, right . . ."; "Really?"

The practice of interactive feedback is crucial to good conversation and conversational approaches to knowing. It allows listeners to test their understanding of what their interlocutor is saying, even as it gives the interlocutor feedback on whether he or she is getting through. It encourages listeners to empathize with the speaker, to use their imagination to "see" what is being said, and to draw on their own experience for analogs and differences. It also allows listeners to steer the conversation in one direction rather than another, even to change the topic entirely.

## Communities

Many Americans see themselves as autonomous, self-directing individuals who, as it were, self-fashion their own lives.[9] A recent national study of teenage religious and spiritual beliefs suggests that this self-understanding is already well established by the time students arrive at college. In a fascinating piece of irony, teenagers are socialized into believing that they are only minimally socialized, fashioned by consumer capitalism and media technology into believing that they are largely self-fashioned, and influenced by the ideals of liberal America into believing that they cannot be unwillingly influenced. This presupposition disguises from teens, and probably also from many of their parents, the real state of affairs.[10]

By situating religion and disciplinary scholarship within their community contexts, this book explores how we're all shaped and formed by our society, culture, and communities, especially our disciplinary community and, if we have one, our religious community. As a historian, I am convinced that society, culture, and key communities define and shape how individuals construe self and world. But in these chapters, I am advancing only a qualified sociological and psychological, not a rigorously epistemological, claim. I do not argue this point as the scholars cited in the endnotes do this for me. I simply assume that we understand better how we construe self and world if we turn our attention to the social and cultural structures, the communities, and the individuals that have shaped and formed us. I believe that our own aptitudes and choices play a role—but only a constrained role—in who we are and what we choose to believe and do. We are free, but far less free than we would prefer or perhaps even expect. If you see the shaping role of society and community differently, I urge you nonetheless to appreciate some instrumental value in my interpretive strategy; it may not be true in some deep philosophical sense, but it can nevertheless help deepen our understanding of the complex interaction between disciplinary and religious conviction.

In this book, the term "community" may apply equally to relatively intimate groups whose members know each other both professionally and personally (such as university or college departments or religious congregations) and to more impersonal groups where many members do not know each other and belong to the community only because of shared expertise and training or shared beliefs and practices (such as national disciplinary communities or religious denominations). For my purposes, it seemed unnecessary to distinguish, as some do, between community and association. Further, I intend the term to be neutral without taking sides in the debate over communitarianism.[11]

## Overview

When we talk about religion and higher education in America, we're largely talking about Christianity. Christianity is America's dominant religion, although its dominance has been waning in recent decades.[12] Still, about three-quarters of the American population self-identifies as Christians and about half of which self-identify as Protestants.[13] The historical material in this book reflects this Protestant Christian preponderance, with occasional references to distinctly Roman Catholic issues. The book seeks, however, to offer an analysis broad enough to serve America's increasing religious and spiritual diversity.

We begin in part I, *Cautions*, with two introductory chapters that illustrate the risks posed by religion on campus. The first chapter, "Cautionary Tales," asks faculty to reflect on the stories our discipline and our college or university may tell about how religion has inhibited intellectual progress, threatened academic freedom, encouraged discrimination, promoted insoluble divisions, and mystified hegemonic power relationships, among other negative consequences. The chapter employs an approach that will recur throughout the book: it calls on faculty to ponder the stories that make up our identity and that inform how we understand our profession and our college or university. Chapter 2, "Encounters," brings the challenge of religion on campus up-to-date and suggests that the accusation that someone is inappropriately proselytizing can cut several ways.

In part II, *Communities*, we explore how faculty are socialized into their disciplines and how in significant ways this socialization has much in common with traditional religious formation. These initial chapters—chapter 3 on religious formation, chapter 4 on disciplinary formation, and chapter 5 on the constraining role of institutional settings such as a college or university—provide a common vocabulary and advance a series of deliberately provocative parallels that inform subsequent chapters.

In part III, *Individuals*, we move to a first-person consideration of how we ourselves entered into disciplinary and institutional communities.

As with "Cautionary Tales," the recommended approach is narrative. "Narrative Identity," chapter 6, explores how the story can be recounted of how we became disciplinary professionals and college or university teachers. In "Inclinations," chapter 7, we tackle how our deep convictions may have inclined our interpretive or explanatory choices as scholars and teachers.

Part IV, *Implications*, begins with "Community Warrant" (chapter 8), which picks up where part II left off, and delves into how communities, especially disciplinary and religious communities, influence scholarly judgments and how such judgments are warranted and justified. This "regulative ideal of a critical community of inquirers"[14] provides the backdrop for chapter 9 on "Academic Freedom." In the American context, academic freedom applies primarily to faculty but depends heavily on the academic freedom of specific colleges or universities. It is a complex notion that also entails responsibilities and significant limitations. It bears on religious conviction in perhaps unexpected ways.

In "Reticence," chapter 10, we return to the issues first raised in "Inclinations," which explored how religious or spiritual convictions may incline scholars to favor one explanation or interpretation over another without ever making an explicit appearance in the scholarly account. We tackle the question of why silence about religious or spiritual influences is often the most prudent policy. We also ask the question of when, if ever, silence might appropriately be breached. Part IV concludes with chapter 11, "In the Classroom," which considers two strategies that faculty may wish to ponder if they are inclined to reintroduce explicitly religious considerations into the classroom: self-disclosure when dealing with moral judgments and natural inclusion when dealing with subject matter on which religious conviction has an obvious, and pedagogically useful, bearing.

Appendices 1 and 2 offer suggestions to seminar leaders and a discussion setup for "Narrative Identity."

Disciplinary professions, the historian Thomas Haskell reminds us, are collective enterprises.[15] Religions are as well. To reflect wisely on the proper role of religious discourse in scholarship and teaching, it helps to recognize how deeply these two communities establish in us fundamental dispositions and controlling habits of mind. It also helps to recognize that our professional discipline is better positioned than most American religious communities to insist that we adjust ourselves to its goods, standards, and practices. It helps, finally, to recognize that disciplinary communities have a good reason for being skeptical about religious discourse in scholarly work or teaching. We need to understand well the communal contexts that have formed and now guide us if we wish to move the conversation about religion on our campuses into the new religiously plural America of the twenty-first century. We start with cautions.

# Part I

# Cautions

# Chapter 1

# Cautionary Tales

In the next several chapters I argue that the professional formation (or socialization) of academics should be seen as a long process of acculturation and induction into a larger, preexisting disciplinary community. This process, like its religious analog, takes the aspiring scholar through supervised stages, punctuated by periods of intense testing and celebratory rites of passage. Like the formation of a professed religious, the professional formation of the scholar inculcates self-discipline, demands an almost ascetic self-denial in service to disciplinary goods and standards, and expects submission to external authority. It demands risk, encourages emotional involvement, and produces strong self-identification with the goals, standards, values, and world construal of the scholarly discipline.

In the process of this professional formation, the aspiring scholar is told stories about the discipline. These stories disclose the values, the history, and the conflicts out of which the discipline arose and against which it often defines itself. These stories are frequently didactic—they convey a moral. As the young scholar becomes increasingly socialized into the discipline and is formed by the practices that define the discipline, these stories become the scholar's own. They become a part of his or her identity.

A similar process occurs at the places where we live out our professional identity. For people who work at a college or a university, the institution's identity takes the form of multiple, intertwined stories. These stories are handed down, passed around, invented, altered, told, retold, and embroidered. They tell of people, events, common practices, triumphs, tragedies, and injustices. Their multiplicity assures that an institution's identity will be thick with complexity, multifaceted, frequently inconsistent and contradictory, and always changing slightly around a relatively stable core.

As we live and work at a college or a university, our own story—our identity—becomes entwined with the institution's stories, taking new form even as it alters (most likely in small ways but sometimes in large) the stories that the members of the institution tell about their community. Institutional stories encourage us not only to do some things in our research and teaching but also to abstain from others.

In this chapter we explore some key narratives that shape how we understand our discipline, our college or university, and the relation of each to religion or at least to specific religious traditions such as the once-dominant forms of American Protestant Christianity. Variants on the more edifying stories may be recounted in introductory textbooks or in presidential addresses at national disciplinary associations. They may be rehearsed in course catalogs and celebrated in fund-raising appeals (although it is sometimes striking how little may actually be said at some colleges and universities about a once-pervasive religious presence). You will probably find such accounts in your national association's web site as well as in that of your institution's.

Less happy tales about the pernicious influence of religion are more likely to be passed on by word of mouth, from older members to newer, or from disciplines more in the line of fire to those more on the periphery. These stories often point to dangers that a responsible discussion of religious motivation in college or university life must seriously address. Religious conviction has done a great deal of harm in American intellectual life and in American colleges and universities. It is unwise to overlook the downside of religious motivation in higher education.

In telling these tales—which I expand on in the chapter "Disciplinary Formation"—I've employed categories from religious studies. If these deliberately provocative parallels seem forced, feel free to replace the religious terminology but do take seriously the didactic moral often conveyed by these stories.

## Disciplinary Myths of Origin

What are the mythic tales that our discipline tells undergraduate and graduate students about the origins of the field? By "mythic" I mean to capture the larger-than-life quality of many of these stories; I do not mean to imply that the stories are false, only that they bear a paradigmatic character that defines the field and invites the apprentice scholar to emulate and admire. These narratives often include tales of epic struggles with those who attempted to compromise the intellectual integrity and autonomy of disciplinary scholarship. Here, we'll focus in on narratives of origin that deal in some way with religion (often the Protestant Christianity of the late nineteenth and early twentieth centuries, but sometimes Christianity more broadly).

## Natural Sciences

During much of the nineteenth century, the status of natural science within America's colleges and universities depended on its close association with Christian theology.[1] Above all, science provided evidence of design. Assuming that God had created the human mind in the image and likeness of the divine Rationality, it was further assumed that the study of science disclosed the rationality with which the Creator had endowed the Creation. "For much of the nineteenth century," historians Jon Roberts and James Turner explain, "natural scientists working in institutions of higher education played a pivotal role, both in the classroom and in their publications, in developing what Theodore Dwight Bozeman has felicitously termed a 'doxological' view of science, that is, the view that the investigation of nature constituted a means of praising God."[2] These natural, easy assumptions about divine Rationality and the Creation gradually allowed scientists to explore nature apart from what the Scripture has to say about nature, laying the groundwork for an eventual separation of the two.

Between 1830 and 1870, scientists increasingly came to limit their discussion to natural phenomena, favoring causal explanations that rested on "secondary causes" rather than on supernatural intervention. With time, the appeals to supernatural explanations diminished and finally disappeared all together. Scientists adopted instead a naturalistic description and increasingly came to believe that supernatural fiat was "not the way in which Nature does business."[3] In effect, what constituted an explanation had changed. If scientists were unable to account for a natural phenomenon, their proper response was not to invoke God but rather to pursue further scientific inquiry. After 1870, most scientists simply assumed that all natural phenomena were amenable to naturalistic description and explanation.

By the last decade of the nineteenth century, the divorce between natural theology and science was complete, with few colleges or universities even offering natural theology courses anymore. The alliance between Christianity and science that had served so well to establish the latter in the curriculum was replaced by a new justification: that science with its rational mode of inquiry provided the surest road to achieving a true understanding of the natural world. The naturalistic assumption had swept the field.

Today's natural sciences first entered the American college and university curriculum as helpmeets to natural theology, providing evidence for design.

> If you're in the natural sciences, what stories hearken back to this transition from handmaiden to serious competitor?

## Social Sciences

About the same time that the naturalistic assumption swept the field in the natural sciences, scholars of what might be termed the "human sciences" broke away from the traditional courses in moral philosophy. These nascent disciplines—history, psychology, political science, economics, sociology, and anthropology—allied themselves with the natural sciences both to acquire some of their allies' prestige and also out of a widely shared conviction that the scientific method provided the surest means to attaining truth.

The timing was significant for the character of these new disciplines. "As disciplines that self-consciously sought to ally themselves with the natural sciences," Roberts and Turner explain, "the human sciences were in a very real sense born with a commitment to methodological naturalism."[4] In contrast to the moral philosophy courses out of which these new disciplines emerged, where providence and divine intervention featured prominently, these new human sciences employed a rhetoric and methodology that was rigorously naturalistic. By the turn of the century, psychologists, sociologists, and anthropologists were employing their scientific methodology to understand religion itself in naturalistic terms.

Concentrating on the discovery of causal relationships and agents, social scientists became resolutely empirical even as they sought to discover "laws" of social behavior. During the Progressive era, they often conjoined their zeal for scientific advance with a conviction that scientific progress would drive social improvement. The prediction and control promised by the scientific method would be put to service for social engineering. By the 1920s, the younger social scientists had embraced the "value free" model of science, equating objectivity with value neutrality. While they continued to see social utility arising from their research, they professed that a value-free approach was crucial to making their results socially useful.

Though administrators continued to insist on character formation as one of the prime goals of higher education, their faculty became progressively more doubtful and resistant to taking on this task. "University administrators' plans to use the biological and social sciences as secular substitutes for religion," the historian of education Julie Reuben explains, "soon came into conflict with younger faculty's conception of their disciplines":

> Academic scientists coming of age in the early twentieth century rejected the utopian visions of science and the ideal of the unity of truth that had been so important to their predecessors. They embraced specialization and rejected efforts to synthesize all knowledge. They began to see the interests of

> their disciplines in a model of science that stressed the importance of factual description rather than constructive adaptation to the environment and that associated objectivity with the rejection of moral values. In adopting this new conception of science, faculty defined their role in the university as producing research and providing specialized training. This more limited role gave scientists more autonomy and freedom from administrative supervision.[5]

While social scientists were in the business of developing a "scientific morality," administrators (and other outside authorities with an interest in the morals of the student body) had seeming justification to intervene when individual social scientists failed in some way to do their proper duty. But once social science was "value free," such intervention was no longer justified. Professional independence was purchased, in part, by denying the disciplines' direct relevance for moral decision making. The distinction between means and ends—with the social scientists discovering the means and others determining the ends—served this transition well.

> If you're in the social sciences, what stories hearken back to the time when your discipline surrendered the task of moral formation and embraced a more "value-neutral" approach?

## Humanities

The humanities grew out of the classical education offered in Antebellum America. Drawing on a long history of liberal education and its (often conflicting) functions,[6] the predecessor courses to today's humanities courses sought to promote a broadly "Christian" morality and shape character by requiring students to read the classics, which were seen as "edifying" texts. In the early decades of the twentieth century, this function underwent revision as humanists (under the influence of especially Charles Eliot Norton[7]) came to champion a different route to character formation. They now sought to form character by exposing undergraduates to the art, literature, and thought of Western civilization. Recent engagements in the "culture wars" and "the battle over the Canon" draw, in an often confused and ironic way, from these earlier developments.[8]

> If you are in the humanities, what stories, recent or distant, laud or vilify the Western "Canon" and its assigned task to promote "American" or "Western" values?[9]

## Exemplary Ancestors

"Myths of origin" within disciplinary fields often include "exemplary ancestors" who embody the attitudes, virtues, and gifts that the field values and hopes young scholars will emulate. They are often heroes in the struggle with outside powers who seek to compromise the field's integrity. Two obvious examples from the natural sciences are Galileo and Darwin, each of whom had a direct run-in with the religious authorities of their day. In the social sciences, the "exemplary ancestors," such as Max Weber, Sigmund Freud, and Adam Smith, also stand for the complicated, often contested relations between the new fields—sociology, psychology, and economics—and varieties of traditional Western Christianity. For decades, the practitioners of the new social sciences often saw themselves in a crusade against the forces of religious obscurantism. They were offering a "scientific" approach to issues that were traditionally dealt with in religious terms. By 1930, the leading social scientists were championing "value neutrality," and had thereby "rejected the notion that their disciplines should provide the college curriculum with unity and moral purpose."[10] This ideal provided the rallying point for much of the rest of the century and is still championed by some stalwarts in the first decade of the twenty-first century.

Who are the exemplary ancestors within your discipline? Do their stories have any bearing on how your discipline thinks about religion?

## Insoluble Disagreement, Subjectivity, and Violence

The Enlightenment arose in part to provide an alternative to the seemingly insoluble religious disagreement and violence occasioned by the Protestant Reformation. During the sixteenth and seventeenth centuries, religious wars wrecked havoc in the Holy Roman Empire, France, and England. Since conflicting claims could not be settled rationally, attempts were made to settle matters with the sword. Out of the ashes of futile and bloody conflict arose a nation-spanning attempt to establish a universal, rational basis for deciding matters pertaining to the common good.[11] The Enlightenment championed "universal reason" as an alternative to "sectarian faith," and the modern American college and university often saw the choice in similar terms. Or so the story goes.[12]

Violence inspired by religion did not end with the Enlightenment, of course. In today's world, religion offers aid and comfort to violence around the globe, even in the United States. The bloody battles in the Middle East between Muslims and Jews; in Northern Ireland between Catholics and Protestants; in India among Hindus, Muslims, and Sikhs; in Sri Lanka between Hindus and Buddhists; in the Balkans among Catholics, Orthodox, and Muslims all illustrate that passionate religious conviction can easily fuel hatred and violence—even in the modern era.

The growing aversion to religion within the American academy during the earlier decades of the twentieth century may, however, stem less from fear of actual violence and more from a distaste for discord and argument over issues (often moral issues) that were seemingly beyond rational adjudication. Worse yet, moral aims, scholars came to be convinced, would contaminate scientific or other forms of "objective" research. This is the story that Julie Reuben tells for the early decades of the twentieth century. By the 1920s, Reuben explains:

> Many of the younger generation of scholars thought that eliminating ethical concerns was the key to achieving scientific rigor and intellectual consensus. These scholars viewed morality as a matter of personal preference. They argued that ethics contaminated scientific research by confusing subjective values with objective facts . . . According to this view, one of the main reasons why social scientists did not agree on the results of their research was that moral concerns colored their interpretation of facts; similarly, moral aims had undermined the research of their predecessors.[13]

In such tales, religious conviction threatens and subverts the peaceful quest for (objective) knowledge and understanding.

> In your discipline's self-understanding, has organized religion ever been seen as a competitor or rival in the pursuit of knowledge and understanding? Have there been actual clashes, and if so, how were they resolved?

## Religious Discrimination in American Higher Education

American higher education has an unhappy history of one religious group—normally "established" liberal Protestant Christianity[14]—discriminating against various other groups, most commonly Jews, Catholics, and various more conservative, often ethnic Protestant denominations.[15]

Many Eastern elite universities and colleges maintained quotas on the admission of Jews well into the twentieth century.[16] By one tally, Jewish students made up nearly 10 percent of over a hundred institutions in 1918–19 while constituting only 3.5 percent of the American population.[17] These percentages would have been much higher, however, had only criteria of intellectual merit prevailed. Anti-Semitism, American nativism, and White Anglo-Saxon Protestant (WASP) snobbishness combined to maintain quotas at this 10 percent level at the elite Ivies. The rationale behind this policy was bluntly stated by the then Yale president Charles Seymour. Here's historian Dan Orien's summary of a lengthy letter that Seymour wrote in response to a complaint by Yale alumnus Leonard Shiman (B.A. 1924):

> Seymour began by denying that Yale excluded any racial or religious group, but he insisted that it was "a definite policy to maintain a balanced undergraduate population in so far as this can be achieved without detriment to the average quality of the student body." If necessary, Seymour continued, the policy might "involve some temporary restriction on the numbers selected from one or another of the nation's population groups in order to prevent distortion of the balanced character of the student body." In defense of restrictive policies, Seymour noted that many "of my Jewish friends have told me that it is because of this balance that they want their boys to come to Yale." Echoing President Lowell of Harvard, Seymour justified quotas as a method of preventing "prejudice against any minority or racial group." He concluded, therefore, that because "the percentage of applicants from Jewish homes was larger than ever before," Yale had "decided to stand by its policy of selective admission and to preserve as in past years the balanced character of the Freshman Class."[18]

Given such a logic for justifying quotas for Jewish admissions throughout the first half of the twentieth century, it is no wonder that some Jewish groups continue to look with suspicion on all quota systems operating in college or university admissions, whether ceilings such as that employed by Yale to limit Jewish admissions or floors intended to promote affirmative action for African Americans.

The underrepresentation of Catholics and conservative, often ethnic, Protestants at the major elite universities through mid-century may be due more to class differences and sectarian alternatives than to overt discrimination.[19] The Eastern elite universities represented "national culture" (which was assumed, without much critical reflection, to be liberal Protestantism). Those ethnic groups that wished to maintain their own cultural identity—which often had a strong religious component—chose instead to found and patronize their own colleges and universities. This enclave strategy was followed by Roman Catholics, Lutherans, and various Reformed denominations.[20] After World War II, church-related colleges in the Protestant tradition began to join the academic mainstream, and Roman Catholic colleges and universities followed suit after Vatican II—developments that have been the occasion for celebration in some quarters and lament in others.

Even after explicit quotas used against Jewish and Catholic students were lifted, bars to professorial positions lingered for some time (and may still exist in some institutions). For example, there was only one Jewish professor in all of Yale College in 1950 and only a scattering of Jews within the professional schools. But, as Orien has shown, over the next two decades, the bars to Jewish scholars came down dramatically. By 1970, 18 percent of the professors in the College were Jewish.[21] This pattern was repeated in the major elite research universities throughout the country.[22] The old Protestant hegemony had come undone, and its undoing benefited first Jews and then Catholics, women, African Americans, and others. Needless to say, scholars from this transitional generation and their students have ample reason to look skeptically on overt religious considerations in higher education, for it was such considerations that kept them on the outside for so many years.

Run-of-the-mill anti-Semitism and American nativism no doubt played a substantial part in the establishment of quotas for the admission of students and in barring Jews, Catholics, and other religious minorities from professorial positions. But we academics are masters at rationalizing our beliefs and so prejudices were often given intellectual rationale.

When the social scientists took on for a time the task of moral formation, they questioned whether Jews or Catholics were capable of conveying what amounts to Christian moral precepts (shorn, to be sure, of their explicitly Christian marks). When the humanities took up the burden of "character formation" after the social scientists had laid it down, they inherited the suspicion that "outsiders" were less likely than "insiders" to appropriately pass on the tradition the West had inherited.[23] By the 1930s, professors in the humanities, as the self-described bearers of "Western culture" (which was often taken to be synonymous with Protestant Christianity), were often the most resistant to the entrance of religious and racial minorities into their ranks.

Catholics were still suspect at mid-century because they allegedly owed allegiance to what many liberal Protestants and secularists saw as a dogmatic, authoritarian, and "un-American" faith. Catholic "authoritarianism," it was charged, was antithetical to the ideals of democracy and free inquiry.[24] Jews were disqualified, on the other hand, because Western civilization and its values depended, the WASP academic establishment claimed, on understanding the New Testament; something they thought Jews lacked the background to do.[25]

Understanding these rationales—and their hollow core—may help us to see why religious perspectives in higher education were rightly marginalized and what has to be avoided if circumspect religious considerations are to be reintroduced in a way that respects and honors our religiously pluralistic society.

> What stories of discrimination are told in your discipline, in your college or university, and in the other communities you inhabit? What do they suggest about the risks of (re)introducing religious discourse into scholarship and teaching?

We come to understand self and world in no small part by the stories we hear and tell. In this chapter we have attempted to surface the stories that rightly caution us to be careful about mixing religious conviction with liberal higher education. If we are to overcome the injustices and mistakes of the past, we must understand what they were, ponder how they occurred, and think carefully about how we might avoid repeating them. The cautionary tales that we learn from our disciplines and places of scholarship and teaching put us on notice. To be sure, the world we now inhabit has changed significantly from the world that gave rise to these stories. But religion is still a potent force in the world and an abiding (although often hidden) presence in the lives of many scholars and teachers. In the next chapter, "Encounters," we take up some cautions rooted in present circumstances.

# Chapter 2

# Encounters

Whether it's Jehovah's Witnesses ringing our doorbell with the *Watchtower* in hand or an intense young man asking whether we've accepted Jesus as our personal Savior, most of us have been on the receiving end of an attempt at proselytization. And while we may understand why they are doing what they are doing, and even applaud their commitment and bravery in making the approach, we are commonly at least a bit uncomfortable and wish they would simply leave us alone. Why is this? And more to the point, why is this especially true within the academy?

To get at the issues, it helps to distinguish among (1) simple disclosures of religious affiliation, (2) sharing one's beliefs without intention of converting the other, (3) witnessing to a religious belief or practice but without attempt at proselytization, and (4) proselytizing itself that includes disclosure, sharing, and witnessing but aims ultimately at converting the other to one's own position.

Let's start with a simple disclosure. "I'm a Lutheran," announces Erik Johnson to little surprise, given his name. Or comments are made in passing that alert the hearer to a religious affiliation, "I ran into Mary at Saturday Mass, and she told me about the faculty meeting" or "We're getting ready for Joan's Bat mitzvah." In the past, faculty (especially from minority traditions) might have been cautious even about disclosure. Recall the discrimination that Jews and Catholics experienced at Ivy League institutions not so many years past.[1] Even today, some church-related colleges and universities have fewer Jewish faculty than their percentage in academe as a whole would lead one to expect, which can lead to understandable sensitivities. And the situation of today's Moslem or Hindu faculty members may well bear resemblance to patterns of discrimination found in earlier decades against Jews and Catholics. But so long as we do not forget the unhappy practices of the past and remain vigilant about continuing these in the present, simple disclosure within an academic context has become largely unobjectionable, although the more militantly secular of colleagues may raise an eyebrow and ask, or at least think, "how can an obviously intelligent chemist—or historian or psychologist; whatever—still believe in that stuff?!" But, to repeat, simple disclosure occurs now, all the time, with little fuss.

In our more openly diverse society, we're also more comfortable than our forebears were in simply sharing our religious beliefs and practices, especially when asked. A Hindu colleague may invite his or her fellow colleagues to a Diwali Festival (the "festival of lights") and explain to those who come how in their part of India the festival honors Lakshmi, the goddess of good fortune. Myths will be recounted; children of the family may do traditional dances; all will be lightly explained. Or Reformed Jewish colleagues may invite a departmental colleague or two to join them for a Seder meal. The five foods on the Seder plate will be explained and how they recall the Israelites' struggle throughout the centuries. When asked by a curious Baptist colleague, an Episcopalian or Catholic may explain the parts of the Eucharistic liturgy that are unfamiliar to a person who has grown up with a simpler service centering on the sermon. Such experiences and conversations help colleagues understand each other better and are often signs of friendship and respect. They allow all involved to deepen their understanding of others without having to resolve or even address who's right and who's wrong—or whether the question of who's right or who's wrong makes any sense.

Witnessing can be both verbal or nonverbal. Consider, for example, dress and jewelry. An Orthodox Jewish male student may wear a yarmulke or a Moslem female student a headscarf, and Christian students of both sexes may wear crosses. They come to class so accoutered and thereby make a statement of allegiance without uttering any words. Some suggest more commitment than others. Little can be inferred from the choice of wearing a necklace cross, except for some unspecified Christian commitment. Wearing a yarmulke or headscarf suggests greater commitment, and may even suggest a more conservative or traditional affiliation. Apparel having religious significance may be so casual as to represent little more than a simple disclosure as discussed earlier, or it might serve as preludes or at least invitations for further conversation, even small talk that may lead to attempts at conversion.

A simple, rather nondemanding form of verbal witnessing often occurs in classrooms. "I am willing to learn evolutionary theory," announces a biology student, "but I believe that a Creator was still involved." "I think that abortion is never an acceptable option," a student declares in a literature class dealing with Maya Angelou's *I Know Why the Caged Bird Sings*. "Torah teaches that the Almighty promises all the land from the Jordan River to the Mediterranean Sea to the descendents of Abraham, Isaac, and Jacob as a perpetual possession," a student asserts in a history class on the Middle East from World War I to the Yom Kippur War, "Judea and Samaria belong to the Land of Israel and should never be given away." Often with witnessing of this sort, the declaration suffices. The student does not feel the need to argue her point; she just wants to be clear about her belief. The biology student may still ace the test on evolutionary theory. The literature student may still write a sensitive essay on how Maya Angelou first saw her pregnancy as distinctly negative and later as a "blessed event." The history student may still be able to handle fairly the various perspectives on the contested status of Jerusalem after 1948. The students feel the need to make a declaration of principle, a statement about who they are and where they stand regarding an issue of great importance to them. But they rest content with the statement and don't attempt to convince anyone else or, unless challenged, to explain or justify their views. It is simple witnessing.

At some point, however, being a simple witness becomes the first step in a journey toward attempted conversion. We turn now to proselytization.

## Proselytizers and Proselytees

In his introduction to *Sharing the Book: Religious Perspectives on the Rights and Wrongs of Proselytism*,[2] the historian Martin Marty distinguishes usefully between *proselytizers* and *proselytizees*. The former does the proselytizing and the latter gets proselytized—although sometimes the roles may be interchanged in the process of proselytizing, as we will see when we explore a bit. In popular imagination, the metaphor of predator and prey comes to mind. "In the dramas that follow, the proselytizer is in one place, specifically, in a religious community or situation," explains Marty.

> This person approaches another with the intent to convert, to help or make this proselytizee come to the same place. In negative imagery that usually colors such incidents, one is to picture the proselytizer assessing the scene, spotting a potential proselytizee, stalking her, watching for a weak moment, getting poised, and then pouncing.
>
> Meanwhile the proselytizee, on the point of being approached, has been occupying a different place than does the one who would convert her. She has her scene, becomes wary of being overtaken, does her own measuring of the distance between her and the proselytizer, tries to stay strong but may grow vulnerable, and then gets confronted.[3]

When seen this way, there is, as Marty observes, a "shadow of violence" that shrouds the encounter. At the very least, it suggests a morally ambiguous situation. Although these associations may not be entirely fair—the predator does not have the good interests of its prey at heart, while the proselytizer usually believes that he is helping the proselytizee achieve a better state—they do reflect the range of uncomfortable feeling we've all experienced when confronted by someone who seeks to convert us on these most intimate issues of belief and practice. In a national survey, for example, about half of the nonconservative Protestants who had been proselytized by an evangelical Christian characterized the experience as "negative."[4] I propose to unpack the notion of proselytizing when applied to the academic setting to give us some conceptual distinctions that may help us think more clearly about the matter.

## Why Proselytizing Is Not Welcomed in the Academy

Martin Marty suggests at least four reasons why we Americans tend to have at best mixed feelings about attempts at proselytization.

First, we like to think of ourselves as a society in which individuals are free to make their own choices. We should have the right, we feel, to choose our own opinions, beliefs, creeds, practices, or parties. Further, we should have the right to be left alone in our free choice of opinions, beliefs, creeds, practices, or parties. Someone else should not be infringing on this freedom.[5] Someone else shouldn't be trying to do the choosing for us. We in the American academy also claim the right freely to choose what we believe, do, and with whom we affiliate, and we often assert this right under the doctrine of academic freedom. These choices seem to lie at the heart of personal identity in America, both within the academy and within the larger society.[6]

Second, we see ourselves as living in a pluralistic age in which, as Marty aptly puts it, "differing peoples with differing opinions, beliefs, creeds, practices, or parties can coexist creatively, or at least neutrally." The proselytizer crosses these boundaries with, we fear, the aim of eliminating difference. She explicitly challenges the communities and traditions that have formed the proselytizee, argues that they are inferior to the alternative she is offering, and exhorts the proselytizee to cross over. She implicitly (and often explicitly) rejects the notion that difference is good and diversity salutary.

This implicit disdain or even hostility toward the "inferior other" naturally arouses fears and resentments among those being proselytized. This is especially true for members of minority communities when the proselytizer represents a majority (or an aspiring majority) group. In the contemporary academy, which sees diversity as a signal good, this implicit drive toward homogeneity can seem particularly repugnant and threatening. We in the academy are committed in theory (if not always in practice) to the principle that understanding, knowledge, and even truth arise from the never-ending, free clash of opinion, evidence, and argument. Unanimity represents a dead stasis. Imposed unanimity represents a fearful tyranny, inimical to the cherished principles of academic freedom. So the goals of the proselytizer may strike some academics as being starkly antithetical to the value of diversity and unremitting questioning that make the academy humanly worthwhile. Of course, some critics of the academy argue that we are committed to diversity in every aspect of life *except* for religious conviction (and political conservatism). There is enough color to this charge that it needs to be taken seriously, if only to be able to explain to critics why the disciplinary rules of the game work as they do.

Third, Marty reminds us that we live in an age where identities are insecure. We Americans are a people much occupied with self-fashioning. In our consumerist society, and under the impact of all forms of media, we are constantly being challenged to rethink ourselves and to choose our (often shifting) identity from the many beguiling elements and possibilities that the commercial and informational marketplace offer. This is especially true for adolescents and young adults, which comprise a substantial portion of our college or university student body.

It is almost a cliché to say that the college years are a time for young adults to wrestle with questions of identity. Given this state of affairs, we may feel obliged to protect our students from religious proselytization at a time when they are particularly vulnerable. In pondering this obligation, however, we need once again to recognize that some outside the academy (and even some from within) see *us* as the proselytizers. They see us as proselytizers for what they term a "secular, naturalistic, and relativistic" understanding of self and world. Whatever the cogency of this argument—and it has enough anecdotal support that critics will not be easily dissuaded[7]—the distinction between education and proselytization may not always be easily drawn to everyone's satisfaction.

Finally, the practice of religious proselytizing often pits absolute truths against provisional claims, comfortable answers against disquieting questions, a habit of certainty against a reflex of doubt. The academy does not take well to absolute claims. On the contrary, it sees itself at least as much in the business of asking questions as in providing answers, and the answers are provisional, as the history of even the hardest of sciences, such as physics, have amply demonstrated. From the academy's perspective, the proselytizer exhibits the mental habits of a dogmatist: this is the absolute, unquestionable truth of matters. But academics are called to question dogmatic claims, including even (or perhaps especially) those within one's own disciplinary field. The two habits of mind seem diametrically opposed.

## Some Ironies

The proselytizer employs doubt as a means to make way for his or her replacement certainties.[8] Conversion often first requires the questioning of the proselytizee's current beliefs and practices, even to the point of completely undermining them, in order to prepare the proselytizee for conversion to the proffered alternative. More ironic still, the act of proselytizing itself and its deployment of doubt makes it more difficult than ever for the proselytizee either to retreat into the once comfortable certainties of his or her tradition or to enter into the new tradition with a naiveté of those who have never experienced real choice. As Marty explains, you can no longer be a naïve traditionalist once the proselytizer has awakened doubt:

> Are you sure that your elders and peers have taught you the truth? Are you sure that you might not better yourself by converting? Here, says the proselytizer, is my alternative. Here, says a competitor, is another option. There are many more. Uncertain, bewildered, the prospect is overwhelmed by relativism and loses the sense of integrity that opinions, beliefs, creeds, and parties must command. Or he must be confirmed in his own absolutism so that he can ward off challenges and seductions.[9]

This choice gives rise to considerable cognitive dissonance, so it is no wonder that we would like the proselytizer to go away and stay away.

There is a further irony in the act of proselytizing, especially in the academy. There is a real risk that the proselytizer, in the act of proselytizing, ends up becoming the proselytizee. The academy offers powerfully convincing ways of understanding self and world. Naturalism, for example, has behind it the cachet of successful science and technology. Relativism makes considerable sense of the great variety of human societies, moral codes, and ways of construing self and world. Secularism offers itself as a plausible "neutral" arena in which the passions of religious commitment might be cooled by the even-handed application of reason.

Those communities that send proselytizers into the academy are fully aware that the tables may be turned. So there is an extensive literature aimed at immunizing these young proselytizers against the academy's alternative seductions. With titles like *The Survival Guide for Christians on Campus* and *How to Stay Christian in College*,[10] these handbooks deal with issues that young college students are likely to face: naturalism, postmodernism, "relativist myths," "liberal myths," and "conservative myths." They offer advice on "developing a Christian mind," "holding your own [in class] without being a jerk," and "dealing with hostile teachers."[11] They also offer advice on other relevant topics, such as whether and how to convert people; how to maintain purity in a world of casual sex, drink, and drugs; and how to avoid being proselytized by groups other than one's own. At their best, these books ask their readers to be thoughtful and responsible about their beliefs and practices as they engage with a new and sometimes threatening experience of higher education.[12] At their worst, they attempt to shore up an anti-intellectual posture that will ill-serve students and frustrate most attempts at real education.

What these authors recognize, and faculty members need to remember, is that faculty have greater opportunity to change the minds of students than students have at changing faculties'. Faculty command the higher ground and considerable authority in any interaction. While it is rare for faculty to engage in explicitly religious proselytization, most campuses have at least one or two faculty who are well known for their aggressive political advocacy in the classroom. We also tend to have our share of skeptics who have no tolerance for religious beliefs of any kind and take frequent opportunity in class to say so. Outside observers profess difficulty in understanding why one practice might be tolerated or even celebrated and another considered beyond the pale.

## Who Has the Advantage?

Some critics see us faculty as the truly dangerous proselytizers. They argue that we in the academy are committed to achieving diversity in every aspect of life except for religious conviction and political belief. They charge us with indoctrinating the young in a "secular, naturalistic, and relativistic" worldview. They complain that we ridicule student beliefs and refuse even to acknowledge "dissenting viewpoints" on what are termed "unsettled questions."[13]

These accusations reveal a dilemma. To be successful teachers, we need to be passionate advocates for the way we together with our discipline understand our subject matter. We are the experts who have undergone years of professional education and demonstrated our mastery to our disciplinary peers. We owe it to our students to give our considered opinions on matters under study. We are obliged to criticize interpretations that we believe to be faulty or without merit. We have a responsibility to push our students beyond their comfortable assumptions, to force them to look hard at unexamined beliefs in the hope that they might grow in understanding.

But what do we do when the tactics of engaged education are taken by students or others to be no more than proselytization for a point of view with which they disagree? What do we say when a challenge to student beliefs designed to make the student think, especially when touching on religious or political convictions, may be stigmatized as a violation of the student's academic freedom "to take reasoned exception to the data or views offered in any course of study and to reserve judgment about matters of opinion"?[14] If we in the academy are unwilling to think through the issues and develop our considered answer to such challenges, outside forces will attempt to impose their own solutions with serious consequences for the academic enterprise.[15] We not only take real risks in opening discussions about the appropriate role of religious perspectives on campus, but we also take real risks if we choose to ban such discussions entirely. We take up some of these issues again toward the close of the book.

# Part II

# Communities

# Chapter 3

# Religious Formation

Entry into a religious community is more like learning a language than acquiring a set of beliefs, more like acquiring a culture through total immersion and practice than mastering a set of rules or propositions. To help make sense of this learning process, I first offer a cultural–linguistic and communal model of religion. Since many Americans today consider themselves "spiritual but not religious," I next supplement the cultural–linguistic model of religion with some comments about spiritual practices. To give content to a cultural–linguistic understanding of religious traditions, I then offer, in the third section, some broad generalizations about religious construals of the cosmos, morality, revelation, and anthropology. With these preliminaries out of the way, I suggest how people are traditionally socialized into a religious community.

It is important to understand that I am offering an *ideal* characterization of *traditional* religious formation. In today's postmodern America, a religious or spiritual journey often takes a more complicated route. Rather than asking "how do I find the best way for me to conform to my community?", we frequently wrestle with the question "how do I choose among the plethora of religious offerings available within my society?" In the concluding section, I briefly describe this postmodern state of affairs.

I recognize that few Americans today experience anything quite so coherent and compelling as traditional religious formation. But it is still worth the effort to understand how the traditional process works. If a case can be made that there is a traditional and a postmodern model for religious or spiritual formation—characterized (or caricatured) as "How do I conform?" (traditional) versus "How do I choose?" (postmodern)—there is arguably still only one quintessentially modern[1] way in which the great majority of academics are socialized into their discipline. And it possesses a striking similarity with traditional religious formation. This similarity is key to appreciating the power and depth of professional disciplinary formation and to grasping its implications for constructive conversations about religion on campus.

## Religion

The historian of religions William Paden rightly suggests that it should give us pause that the term *religion* is "completely equivocal" and has no agreed-upon definition, and that some cultures do not even have a term that corresponds to the Western generic *religion*.[2] Yet some rough definition is necessary if we're going to be able to generalize at all.[3]

For his part, Paden characterizes religion as "a system of language and practice that organizes the world in terms of what is deemed sacred." In this characterization, religion is more something people *do* than something they *believe*, although both practice and belief play their part. The key to what makes behavior religious is that it takes place "with reference to things that are *sacred*." Regardless of how *sacred* is defined, that which is sacred for its adherents is "always something of extraordinary power and reality."[4] Summing up, Paden observes, "Whatever else religion may be said to be, it is at least a form of human behavior and language, a way of living in the world, and can be studied as such."[5]

Although appropriately cautious, when pressed, Paden likens religion to a system of language and practice. This suggestion comports reasonably well with the Yale theologian George Lindbeck's more elaborate model of religion as a cultural–linguistic system that "shapes the entirety of life and thought." Religion, according to Lindbeck, "is similar to an idiom that makes possible the description of realities, the formulation of beliefs, and the experiencing of inner attitudes, feelings, and sentiments." Lindbeck continues:

> Like a culture or language, it is a communal phenomenon that shapes the subjectivities of individuals rather than being primarily a manifestation of those subjectivities. It comprises a vocabulary of discursive and nondiscursive symbols together with a distinctive logic or grammar in terms of which this vocabulary can be meaningfully deployed.[6]

A religious tradition, according to Lindbeck, is correlated with a form of life and has both cognitive and behavioral dimensions. "Its doctrines, cosmic stories or myths, and ethical directives are integrally related to the rituals it practices, the sentiments or experiences it evokes, the actions it recommends, and the institutional forms it develops."[7] Lindbeck's model of religion emphasizes how human experience "is shaped, molded, and in a sense constituted by cultural and linguistic forms":

> There are numberless thoughts we cannot think, sentiments we cannot have, and realities we cannot perceive unless we learn to use the appropriate symbol systems. . . . A religion is above all an external word, a *verbum externum*, that molds and shapes the self and its world, rather than an expression or thematization of a preexisting self or of preconceptual experience.[8]

According to this model, the religious cultural–linguistic system is prior to and shapes the experience, emotions, sensibilities, and beliefs of the community. It strongly influences how the members of the community construe self and world.[9]

Of course, the relationship between religion and experience is dialectical. The cultural–linguistic system that is a religion may be prior to, and shaping of, experience, but experiences alien to the system can reshape the system in turn. Lindbeck offers the example of how the "warrior passions" of the barbarian Germans and Japanese transformed pacifistic Christianity and Buddhism—and were changed in turn. Yet the religious system was the "leading partner" in this dialectical interaction.[10]

Less technically and with an eye to its cognitive aspects, Lindbeck describes religions as "comprehensive interpretive schemes, usually embodied in myths or narratives and heavily ritualized, which structure human experience and understanding of self and world."[11] They serve to describe and explain that which is taken to be maximally important.[12] Though not perfect, Lindbeck's characterization captures the aspects of religion most relevant to this book.

The Harvard theologian Ronald Thiemann has extended Lindbeck's model to bring out the communal and public nature of religious faith. Faith, Thiemann explains, "is not primarily an individual phenomenon; it is, rather, an aspect of the life, practice, and world view of a religious community."[13] Drawing on Lindbeck's shorthand description of religions as "comprehensive interpretive schemes,"[14] Thiemann argues that comprehensive schemes are located in communities and provide the community with its identity:

> As comprehensive schemes, religions seek to interpret the whole of reality with reference to the fundamental core of convictions, narratives, myths, and rituals that establish the identity of a community. Properly understood, faith is the set of convictions that defines the identity of a community and its members. Those convictions do not dwell in some private inaccessible realm; they are present "in, with and under" the myths, narratives, rituals, and doctrines of the community. If you seek to understand the faith of a religious community, you must inquire into its literature, lore, and liturgy.[15]

We shall explore some implications of the communal nature of religion presently.

The cultural–linguistic system that is a religion has a history, and within that history the system has changed under the stimulus of changing circumstances and contending construals of the beliefs and practices deemed proper to and definitive of "true faith and practice."

## Spirituality

Many people today, it seems, prefer to describe themselves as "spiritual" rather than "religious."[16] What might this mean? For many, spirituality is an alternative to organized, institutional religion. "I am spiritual but not religious," a person says, meaning that she professes certain spiritual beliefs or engages in particular spiritual practices but does not belong to a church or synagogue or mosque.[17] For these Americans, *religion* has come to mean *organized* religion, and their claim to be spiritual but not religious means that they feel related to a divine or supernatural or transcendent reality apart from organized or institutional religion. Some, including recently surveyed undergraduates, seem even to be "constructing their spirituality without much regard to the boundaries dividing religious denominations, traditions, or organizations."[18] They may even be borrowing bits and pieces—the "bicolating" of spirituality, as Tom Beaudoin describes it[19]—from a variety of different religious traditions. Wade Clark Roof and Tom Beaudoin single out, respectively, baby boomers and Generation Xers for this new marketplace approach to spirituality.[20]

But this anti-institutional sense is not the only possibility, for spirituality can also take an institutional form. We see this most clearly when we shift our focus to the means by which spirituality is expressed. Spiritual practices include devotional practices, such as prayer or meditation; practices aimed at enriching a person's spiritual life, such as reading spiritual literature or attending a retreat; practices aimed at expressing one's spirituality, such as singing or art; and practices that derive from one's relation to the sacred, such as hospitality or support for the poor or disenfranchised. One advantage to spiritual practices in the contemporary era is that, as the sociologist Robert Wuthnow observes,

> If exposure to diverse religious teachings erodes the plausibility of particular beliefs, practices appear to be somewhat more resistant to such erosion because they can be defended less as absolute truths and more as interchangeable activities through which people seek a common experience of the sacred.[21]

Spirituality in its various guises can be an effective deployment of *bricolage*—that is, the very human practice of assembling and employing ideational odds and ends.

Obviously these spiritual practices may be exercised within a traditional religious community, and they are often first acquired there through the process of religious formation. But sometimes the option of spirituality entails switching institutional context, from traditional religious communities to institutions such as various "parachurches" or "parasynagogues," or to support groups, such as Twelve-Step programs.[22] And sometimes the option of spirituality is seen as an alternative to religion (at least to religion understood as an institution with prescribed beliefs and practices). This noninstitutional form of spirituality involves the spiritual practitioner in various spiritual practices apart from any traditional institutional context. It often expresses individual choice and is sometimes explained as a way of being true to one's inner or authentic self.[23]

In all these varieties of spirituality, the spiritual practices are often modified, yielding a personal and even individualistic spirituality. In cases where individuals begin their spiritual journey within a traditional religious community, the effects of religious formation will persist even if these individuals leave their traditional communities all together (or switch communities), though often in modified form. Although perhaps shorn of the context that may have lent them their original meaning and cogency, these spiritual practices remain personally meaningful and important in shaping how the person construes self and world.

## Religious or Spiritual Beliefs

To this point I've suggested that religions formally resemble cultural–linguistic systems. Of course, religion is distinguished from other cultural–linguistic systems by the material claims it makes. Religions have content, and that content tends to take the form of specific claims about the cosmos, human beings, human and natural history, ritual, obligation, and belief. Now it is admittedly hard, in anything but a trivial sense, to frame useful generalizations that adequately cover the variety of claims that can be found in the world's major religions. But again, we must do something to inform the argument developed in later chapters.

Since we're going to be contrasting religious (or spiritual) construals with academic alternatives, I propose we confine our consideration to rather general intuitions regarding (1) cosmology and how it is known, (2) morality, (3) "revelation" through sacred scriptures, and (4) anthropology. That is, in very general terms, I invite us to focus on elementary, often only vaguely articulated assumptions about how the world is construed and best understood, what proper human conduct is thought to be, how both these things are known, and, finally, how the nature of human being is understood. And even in these broad categories, I'll offer only generalizations. In thinking these things through yourself, you'll need to supply the content peculiar to the tradition(s) with which you are familiar.

Here are five, I hope useful and not too simplistic, generalizations:

1. Religious traditions embody the conviction (or the experience) that the cosmos itself is grounded in an ultimate, absolute Reality or Truth that sustains and warrants more limited claims about meaning, purpose, and value in the cosmos.

On the basis of this conviction or experience of ultimate, absolute Reality or Truth, religious traditions reach several conclusions:

2. There are truths about the cosmos even if unknown or imperfectly known by human beings. These truths are independent of human construction or projection. In most religious traditions, the nature of this ultimate or absolute Reality or Truth is revealed primarily by contingent, temporally, and physically situated narratives.
3. There are also moral obligations arising from the nature of that Reality. As with the assumed truths grounded in this ultimate Reality, these obligations may be unknown or only imperfectly known. These obligations are also independent of human construction or projection. They, too, tend to be revealed primarily through contingent, temporally and physically situated narratives.
4. That ultimate, absolute Reality is disclosed (or discloses itself) partly through sacred scriptures. These scriptures may advance arguments that are abstract, universal, deductive, and certain in a logically formal sense. But commonly, these scriptures convey their content through the use of contingent, temporally and physically situated narratives. It should be added that these scriptures claim an authority and meaning that is independent of individual human readers. A reader properly enters into a moral relationship with the sacred text in which the text is the principal agent and the reader is expected to conform himself or herself to the text.
5. On the basis of material claims 1 through 4, most religious traditions embrace a view of human beings—their purpose, potential, meaning, and worth—that rejects some claims about human beings as being too exalted and others as being too belittling. To put this another

> way, most religious traditions resist views on human nature or human being that either suggest little or no limits to human potential, freedom, and moral judgment or so reduce human being to impersonal or deterministic forces that human potential, freedom, responsibility, and moral judgment become effectively illusions.

Theologians, ethicists, and others can turn these broad assumptions into carefully crafted propositions, but most religious people do not logically work out their cosmological, moral, revelatory, or anthropological convictions in this way. Rather, when they think about certain issues, broad convictions like these tend to nudge them in some directions rather than others. In later discussions, we'll explore how such general assumptions—religious or disciplinary or both—may act at a deep level, informing our inclinations in religious matters and in our disciplinary work in subtle and profound ways.

## Lay Formation into a Religious Community

Entry into a religious community commonly begins at birth and the child is then "brought up in the faith." Less common but still not unusual, an adult may choose to join a community and then go through a period of socialization and formation to turn himself or herself into a full-fledged member of the religious community. I shall first offer a model that will need to be qualified to fit the contemporary American situation. We'll turn to the qualifications in the chapter's concluding section.

In the drawn-out case of the child, and more abbreviated case of the adult, the entry into full membership into a community requires the gradual internalization of the narratives, myths, rituals, practices, and beliefs that give the religious community its identity. In this process, the initiates—whether children or adults—go through stages in which they observe, undergo instruction, participate first in limited and then later in more expansive ways. As they become more fluent with the language and and more familiar with the culture of the community, they will often be tested and will undergo formal rites of passage. In the course of this socialization, the initiates are expected to become emotionally committed to the community, to construct their core identity around their community membership, and to understand self and world through the community's eyes and language.

Consider two examples: the practice of Christian prayer and the religious reading of texts. (I employ the Christian example because I know it best; I encourage those of you who understand better other traditions to recast this account to fit your experience.) The practice of Christian prayer characterizes and defines the Christian community. Those who are Christians are socialized into the discipline of prayer over many years, both at home and in the worshipping community that is the church. As is most commonly the case with religious traditions, Christians begin this process as children.

At home they learn from their parents and older siblings. They are taught model prayers to say before going to sleep (e.g., "Now I lay me down to sleep, I pray the Lord my soul to keep. See me safely through the night, And wake me with the morning light. Amen") or before meals (e.g., "Come Lord Jesus and be our guest and let thy gifts to us be blessed. Amen"). Even at this elementary "childish" level, they learn the elementary form of a prayer (invocation, body, closing) and some of its special language (e.g., "Lord," "Lord Jesus," "thy," "Amen") and its special concepts (e.g., "Lord Jesus," "soul," "blessing"). They are encouraged to believe in a God who cares for human beings, who controls the world. They are told that human beings have souls. They are introduced to death and to the idea of life after death. In short, even at a very young age, they enter into an acculturating process that shapes how they see themselves, other people, and the world in which they live.

In the church, they first listen and observe what the adults and older children do. They hear prayers of invocation, prayers of thanksgiving, prayers confessing their sins and seeking forgiveness, prayers asking God's intercession for themselves and others. Even as relatively small children, they may be encouraged by their parents to add the name of their sick grandmother when the pastor asks for the congregation to add their concerns to the intercessory prayer. As they learn the Lord's Prayer, they are encouraged to join in its unison recitation in the Eucharist.

Over the span of many years, children gradually acquire the vocabulary, the syntax, and the semantics of Christian prayer within their denomination. They learn that a prayer begins with an address to God (e.g., "God," "Lord God," "Lord Jesus," "Almighty God") and closes with a standard formula (e.g., "In the name of Jesus," "In the name of the Father, the Son, and the Holy Spirit," "through your Son, Jesus Christ our Lord," or simply "Amen"). They learn the "idiom" of prayer, including, perhaps, various archaic forms (e.g., "thou," "thy," "thine"), specialized terms (e.g., "steadfast love," "gracious mercy," "thanksgiving," "kingdom of God"), and common phrases (e.g., "pour out your Holy Spirit," "protect and comfort," "thought, word, and deed").

With time, they come to realize that there are different types of prayers: some expressing thanks, others begging for forgiveness, some asking God's support for themselves and their community, others asking God to look after people around the world whom they don't know. They gain a sense of the proper order of topics—confession precedes assurances of forgiveness and both precede petitions for aid and comfort; thanksgiving comes commonly at the beginning or end—and as they grow older, they come to understand the logic that dictates this ordering.

They also learn the dispositions appropriate to prayer. They see adults standing with their heads, bowed and their hands folded, assuming the posture of prayer. They are urged to emulate their example and are shushed if they interrupt the prayer with questions. In the communal prayer, the whole congregation is urged to align their hearts and minds with the words of the spoken prayer, and the learners are also encouraged to do so. They are urged to confess their faults and shortcomings and are assured that God will forgive them. They are told to bring their problems "to God in prayer." They are reassured that "God loves you."

Much of this comes to them through observation and participation. But they also attend Sunday school where they are challenged to memorize standard prayers including various psalms. Important concepts like "grace" or "steadfast love" or "sin" are explained in simple terms, and stories are told to exemplify what these abstractions actually stand for. They may be asked to offer a prayer at the beginning or end of the Sunday school class, and they'll be praised or corrected for what they produce. With time, they become reasonably fluent in the practice of Christian prayer. And as their tacit and active knowledge increases, they move from peripheral participation toward full participation, from childish to mature practices.

Over the years, they learn not only how to "do" prayer but also how to see the world through a lens partly shaped by the practice of prayer (often with, in our modern society, a strong admixture of doubt). For example, they come to see the world as a place in which God acts, but often in ways that they do not understand. They learn to interpret what an "answer" to prayer might look like and why some prayers may not be answered. They come to see themselves as beings who fall short of what God expects of them, as beings that require forgiveness. They learn to pray for others and thereby care for others. In short, as they are inducted into the practice of Christian prayer, they are also inducted into a way of construing self and world, into a way of seeing, understanding, explaining, feeling, and imagining.

Let's look at the other example: the religious reading of sacred texts.[24] Again, formation into the practice begins commonly in childhood. It starts with listening to others read reverentially from the sacred text or texts. As the child grows, she may be asked to memorize portions of the text. She will have occasion—probably frequent occasions—to hear portions of the text read and expounded on by a pastor, rabbi, or other leaders of the religious community. At some point, she may take part in a group dedicated to studying the text. In this process she will be introduced to the various ways the text is interpreted and the aids the community employs in its interpretation: dictionaries, cross-references, commentaries, and so on. She will first observe and may later herself engage in discussion and debate in which the citation of verses from the sacred text or texts is used to prove or disprove a particular position or interpretation. She may have opportunity to pray through parts of the text. And in almost every worship service, excerpts from the sacred text or texts will play a role in the liturgy.

Again, this process of formation occurs over a period of time and involves observation, instruction, and personal practice. It often calls for a mentor (or mentors) who guides and instructs. It requires, at some point, emotional investment and the cultivation of a situational discrimination that allows the mature practitioner to use the sacred text in ways appropriate to changing circumstances and cases. Finally, the religious reading of sacred texts tends in most traditions to require of the reader respect for and deference to the text itself. Within the context of religious reading, the text has its own authority, independent of the reader's. It makes claims on the reader, including moral claims. The religious reading of a sacred text is not only a practice, it is a *spiritual* practice that significantly defines the religious community itself.

As will be argued in the next chapter, this religious way of approaching a sacred text differs markedly from academic reading, especially with regard to the reading of sacred texts.

If you harbor any doubt that members of a religious community are extensively formed into the practice of religious reading, consider how much time and energy the leaders of religious communities spend instructing their members on the meaning, the importance, and the crucial relevance of these texts. Consider the special way the texts are handled, housed, and carried about within the community. Religious reading is *reverential*, and the text possesses its own authority apart from any reader.

To sum up, in traditional religious formation, we usually acquire the language and culture of a particular religion while we are still young and particularly impressionable. The process stretches over considerable time, comes with deep emotional valences, and becomes intertwined with our self-identity. As a result, we may find this way of seeing and knowing natural and may take it as self-evident and self-evidently true. This intense socialization and acculturation instills in us an "orientation"[25] to the world, creating strongly held "background" or "control" beliefs that influence how we see, understand, and experience things within our religious community and beyond.

# Formation and Choice in Postmodern America

Few Americans today experience traditional religious formation in the ideal form that I have just described.[26] In contemporary America, unlike earlier eras,[27] religious (or spiritual) formation is rarely so straightforward or simple. As children observe, imitate, and identify themselves with their own religious community, they also observe, imitate, and identify themselves with the various other powerful and competing cultural offerings of the larger society. Even as they learn how to construe self and world through the eyes of their religious community, they also learn how to construe self and world in other ways offered by, say, television and, more recently, the Internet.

Inevitably there will be some tension between major alternative ways of construing self and world. This tension can easily give rise to cognitive dissonance, where the individual feels pressure either to choose one construal and reject all others or to find some (even makeshift) way of reconciling the various competing construals.

Some people resolve this tension by rejecting the various secular construals and embracing what is seen to be a strict, noncompromising religious orthodoxy (or orthopraxy, i.e., right practice). This orthodoxy calls into question all alternatives and offers itself as solely right and true. Such a choice can lead into one form of fundamentalism or another. Academics or future academics rarely take this direction, and even if they do, they rarely stick with it. The tension between their disciplinary standards and construals and the demands of noncompromising religious orthodoxy seldom allows this option. (Religious traditions such as Orthodox Judaism, which emphasizes orthopraxy over orthodoxy, may more easily coexist with disciplinary construals, but there still will be tension.)

Others resolve this tension by rejecting the religious construal of their upbringing (and all other religious construals) in favor of a dominant secular alternative, often some form of materialism or naturalism. This is a common choice for academics or future academics, and for good reason, since it usually aligns one's larger construal of self and world with the assumptions and approaches employed within the discipline.

Finally, many folks resolve, or at least cope with, this tension by a process of accommodation.[28] This is the most common resolution for religious academics. This accommodation can take various forms, and the forms are often combined in idiosyncratic ways. One takes bits and pieces of one construal and combines them with bits and pieces from another.[29] The reconciliation can be as straightforward (but problematic) as a "two-truths" approach or as convoluted (and idiosyncratic) as a Christian physicist's elaborate natural theology. As an example of the former, consider the tack that the paleontologist and popular science writer Stephen Jay Gould employs in *Rocks of Ages: Science and Religion in the Fullness of Life*, where he posits "non-overlapping magisteria," where "magisteria" are domains of authority. Essentially, science and religion deal with different domains and have their own discrete magisterial authority. The natural world is the domain of science, and the moral world that of religion.[30] At the other end of the spectrum, consider the writings of the physicists and theologians John Polkinghorne and Ian Barbour, who attempt to reconcile elementary particle physics with traditional theology (in Polkinghorne's case) or process theology (in Barbour's).[31] And of course one may depart (organized) religion entirely and accommodate competing construals by adopting some form of spirituality.[32] This is the solution of increasing numbers of people in today's world.

Whether religious academics have undergone something resembling the traditional model of religious socialization or some postmodern process of accommodation, their religious socialization is that of a lay person. It is not, in most cases, anything comparable to the rigorous formation undertaken by a professional (i.e., a minister, pastor, monk, or nun). The difference between lay and professional formation needs to be kept in mind as we explore in the next chapter the formation of disciplinary professionals.

# Chapter 4

# Disciplinary Formation

The key element in disciplinary professionalization for us to recognize is it's communal dimension. As historian Thomas Haskell explains in his study of the emergence of professional social science in America, a social theorist's work—and by extension, all scholarly work—is professional to the degree that "it is oriented toward, and integrated with, the work of other inquirers in an ongoing community of inquiry." Disciplinary professions are, as Haskell says, collective enterprises. Their telltale marks include journals, conferences, and more or less uniform expectations for training and certification carried out by disciplinary professionals working within a limited number of recognized research universities. Professionals are crucially oriented toward their professional communities and are responsible to their professional peers. They exercise a unique authority as disciplinary professionals and guard that authority from outside interference or usurpation through vigilance regarding outsiders and strict self-regulation.[1]

An academic discipline can be understood as a community of scholars that seeks to acquire, extend, evaluate, and disseminate systematic knowledge about a particular subject.[2] The disciplinary community establishes and maintains standards of excellence that practitioners aspire to meet in pursuit of the discipline's goals. In the process of achieving its goals according to its standards, an academic discipline extends the ability of human beings to know and understand and to evaluate what they know and understand.[3] The goods, standards, and practices that constitute an academic discipline each have a history, and that history includes debates about what those goods, standards, and practices should be. In a living discipline this debate is ongoing and often contentious. It is a characteristic and arguably a virtue of academic life that doubt and serious questioning tend to find greater acceptance than is accorded in, say, religious communities.

Academic disciplines arose gradually in the late nineteenth and early twentieth centuries out of a process of specialization and differentiation fueled by a confidence in science and an appreciation that advances in understanding and knowledge were the progressive achievement of communities of professional scholars. Let's briefly review the history.[4]

## Antebellum America

Up to the time of the Civil War and for some time thereafter—it varies by discipline and by college—faculty had been primarily oriented toward two communities: the college community where they taught and the religious denomination that sponsored and controlled the college where they taught. Of the two, the religious community tended to play the major role in faculty's formation. But even at this early stage in the development of American higher education, diversity among the Protestant denominations that sponsored, staffed, and attended college had inclined many colleges to adopt an integrating perspective that represented a broad Protestant (and even to some degree, Deist) consensus in place of a particular theological vision specific to the sponsoring denomination.[5] Their academic understanding of self and world was still at bottom religious (normally Protestant Christian of an evangelical sort) but shared among Protestants and Deists of various stripes. They all saw their occupation as a form of ministry or service, intended to form their students in Christian morality and inform them about the harmony between nature and nature's God.

On the eve of the American Civil War, there were approximately 250 colleges in America.[6] Most of these were governed by boards of trustees dominated by ministers, or at least having a substantial ministerial presence. College presidents were normally ordained ministers, and many of the faculty were ministers as well. The curriculum was organized around several key assumptions that had profound religious implications. Even the public institutions tended to have ministers on their boards, required attendance at Protestant chapel services, taught the Bible and Christian doctrines, and expected the education, as a whole, to be broadly Protestant Christian (as understood at the time). The goal of education, at both public and private, schools, was, in no small part, to form unruly and ignorant boys into educated Christian gentlemen.

Most faculty were generalists and saw themselves primarily or exclusively as teachers. Few did research, and even fewer saw research as a (much less *the*) defining element of their vocation as college faculty. Rather they saw their task as passing on the accumulated knowledge of the West and in the process forming Christian gentlemen fit to take their place in (broadly Protestant) Christian society. While a handful had become experts in a specialized area, they were not disciplinary specialists and professionals in the sense that would arise toward the end of the century. They were not yet a part of the disciplined disciplinary communities of inquiry, because such communities were yet to arise.

Their dual membership—in college and church, within a largely Protestant culture—was reflected in the education they offered.[7] College faculty in antebellum America generally operated with several assumptions regarding God, revelation, Christianity, nature, and human beings. At the risk of caricature, let me sketch the outline of these assumptions, acknowledging as I do so that the order of presentation is arbitrary because the assumptions are interrelated and reinforce each other.

It was assumed that God was the creator of the world, including all of nature and human beings. God was also the author of scripture. Nature and scripture must agree with and reinforce each other since they were both made by the same author and creator. Human beings, as creatures of a rational God, had been created with rational faculties that, if properly exercised, would allow them to understand much about nature, themselves, and their creator. These rational faculties were best exercised through the patient gathering and differentiation of facts on the basis of which generalizations could be made to reveal general maxims or laws. This empirical, inductive, "scientific" approach could be applied with equal success and legitimacy to human beings and society as to nature. It could be employed to determine laws of morality as well as laws of physical nature.

These base assumptions are clearly at work in two staples of the antebellum curriculum: natural theology and moral philosophy. Natural theology dealt with that part of theology that contemporaries believed could be known by human reason alone without the aid of revelation. Its authority was built on the assumption that God had created human beings with faculties capable of discerning connections between the creation and its creator. In the antebellum curriculum, natural theology often played a "doxological" role[8]; that is, it allowed men of science to praise God through their study of nature.[9] But in discussing natural theology, faculty only rarely stopped at praise. They also advanced arguments that sought in the minute investigations of nature to find evidence for God's existence and attributes. Natural theology became the prime weapon in the apologetic arsenal for theism and against skepticism.

The label "moral philosophy" may suggest to us a rather specialized course, something like Ethics offered by the Department of Philosophy. But during much of the nineteenth century, the moral philosophy course was the capstone of a collegian's education. It sought to pull together all that a student had learned in the set curriculum of his day and arrange it into a coherent Christian system of knowledge and duties.[10] Moral philosophy had as its purview human motives and obligations, the social relations of human beings, the harmony of nature and Scripture, the agreement of natural science with human morality, and the unity of the true, the good, and the beautiful. As an indication of its crucial integrative function, it was a year-long course offered to seniors and commonly taught by the college's president (who, you may recall, was usually an ordained minister).

Instructors in natural theology and moral philosophy largely took for granted that human beings had been created by God with rational faculties that allowed them to understand the rational world that God had made. The appropriate approach to this understanding of God's works was to rely on the patient gathering and differentiation of facts. On the basis of these accumulated and categorized facts, they could then generalize patterns or laws that would be tested against further facts. It was an empirical and inductive approach that rejected hypotheses or grand theory. It was a species of Baconian inductionism filtered through Scottish Common Sense realism.[11] It was understood to be the only appropriately scientific way to determine not only laws of nature but also laws regarding human beings, society, and morality.

This last point is worth stressing because it may seem strange to us. The discovery of moral laws was thought to be a scientific exercise involving the rational examination of data derived from human experience including human consciousness. Moral laws could be derived in much the same way as could laws of nature. Let's consider an example from Francis Wayland, who was president of Brown University, an ordained Baptist minister, and the most successful textbook writer of his era. In his *The Elements of Moral Science* (1835), he argues (according to George Marsden's account on which I am relying here) that ethics is as much a science as physics, and he terms it "the Science of Moral Law." "Wayland was convinced," Marsden explains, "that the laws of morality, essentially 'sequences connected by our Creator' of rewards and punishments for various acts, could be discovered 'to be just as invariable as an order of sequence in physics.' "

Wayland and other scholars of this period argued that these moral laws could be known by conscience, by natural religion, and by biblical revelation, and he urged the reader to employ all three approaches, recognizing that the moral laws of revelation, while "in perfect harmony" with natural religion, provide facts about matters such as the atonement or the afterlife that we would not know otherwise. "Wayland's method," Marsden continues, "is to 'derive these moral laws from natural or from revealed religion, or from both, as may be most convenient for our purposes.' "[12] And in each approach, whether in examining conscience, studying nature, or reading scripture, the same careful attention was to be paid to the assembling and examining of fact from which maxim or laws were derived.[13]

Instructors in moral philosophy and natural theology also assumed and taught that God was Scripture's Author as well as nature's Creator. Scripture and nature were the two "books of God." Scripture was primary, but the book of nature, it was thought, could also be read to understand God's nature and will. "Nineteenth-century Americans were confident that natural science supported Christianity," Julie Reuben explains, and offers striking illustrations of the point,

> "If God is," wrote Harvard philosopher Andrew P. Peabody, "he must have put his signature on his whole creation no less than his impress on his manifested or written word. The hieroglyphs of nature must needs correspond to the alphabetic writing of revelation." Whenever nature and the Scriptures "give instruction on the same subjects," concurred Francis Wayland, the author of America's most popular moral philosophy text, "they must both teach the same lesson."[14]

Scripture and nature should agree with and reinforce each other. "The truths of revealed religion *harmonize* perfectly with those of natural religion," wrote Wayland (emphasis in the original). "So complete is this coincidence as to afford irrefragable proof that the Bible contains the moral laws of the universe; and hence, that the Author of the universe—that is, of natural religion—is also the author of the Scriptures."[15] The fact that scientists and others in the late eighteenth and much of the nineteenth centuries found these arguments convincing suggests how widespread and unquestioned was the belief regarding God's authorship of nature and its implications for Christian, or at least theistic, belief.[16]

Yet there was an irony hidden in this easy assumption of congruity between the books of Scripture and of nature. On its basis, scientists could in good conscience choose to explain nature solely in empirical and inductive terms with no reference to explicitly Christian or even theistic considerations. They simply assumed that no conflict between Christianity and empirical science could arise. God had created the world and God had authored Scripture. How could there be a conflict? And so, over time, the practice of science (apart from its doxological uses) proceeded apart from religious considerations, saving only, as Marden puts it, "that whatever laws were discovered by this autonomous scientific method must be acknowledged as evidence of the wise design of the Creator." Marsden remarks further that, "as several historians have pointed out, this amounted to a 'rickety compromise' between piety and the ideal of absolutely free scientific inquiry."[17]

Much of this broad Protestant consensus was still operating in 1870, but some subtle and not so subtle changes were occurring that set the stage for a complete reorientation within a generation. Between 1830 and 1870, scientists, who were in the vanguard of this change, came increasingly to specialize on isolable problems that allowed them to focus in on the linkage between effects and putative causes. As a result, they came increasingly to limit their discussion to natural phenomena, favoring causal explanations that rested on "secondary causes" rather than supernatural intervention. With time their appeals to supernatural explanations diminished and finally disappeared all together. Instead, scientists adopted a naturalistic description and came increasingly to believe that supernatural fiat was "not the way in which Nature does business."[18] In effect, the emerging scientific disciplines were gradually developing new standards with regard to what constituted a valid scientific explanation. The community was also coming to expect of its members that when they failed to account for a natural phenomenon, the appropriate response was not to invoke God but to pursue further scientific inquiry. After 1870, most scientists simply assumed that all natural phenomena were ultimately amenable to naturalistic description and explanation.

## Higher Education Reorients

The momentous shift toward professional disciplinary specialization occurred in the period between the end of the Civil War and the onset of World War I. During this period, the nation underwent rapid changes in industrialization and urbanization. Immigrants arrived in huge numbers and greatly complicated the religious and ethnic complexion of America. The Morrill Act of 1862 gave impetus to the founding and development of America's great land-grant institutions. These new foundations opened higher education to many more people, accelerated the offering of professional training, and legitimated education in increasingly technical subjects. Americans who had studied in Germany brought back new notions regarding higher education and research. From 1880 to 1910, the captains of industry partnered with the captains of erudition (as they were called) to undertake the creation of the new university. Some, such as Johns Hopkins University, University of Chicago, and Leland Stanford, Jr. University, were new foundations. Others, such as Harvard, Yale, and Princeton, were colonial colleges that were reformed according to the new style. The historian John Thelin points to the founding of the Association of American Universities (AAU) in 1900, with 14 charter members,[19] and the appearance of Edwin Slosson's *Great American Universities* in 1910[20] as signs of the arrival of the new, research-oriented university. The ideal was grand, its accomplishment still modest in practice; many of America's colleges continued for some time much as before. But when we now look back to this period, we can see in this elite group of universities the new specialized research orientation that would, with time, sweep higher education.

Americans had always harbored great hopes for science, but in these crucial decades, confidence in science and the scientific method (variously understood) soared and knowledge of the world exploded, giving rise to new social arrangements in scholarship. As researchers undertook ever more specialized research, it soon became impossible for any one person to encompass all that was known and all the ways in which such knowledge could be acquired. The enterprise of coming to know the world became necessarily a community project, for only a community could encompass, through its many members, the burgeoning knowledge and the manifold techniques and approaches required to advance knowledge. Specialization, especially in the natural sciences, could be found decades earlier. But now specialists in a range of social sciences and humanities also began banding together in order to master their ever-expanding fields. These disciplinary communities each had a distinct subject matter and distinguishing methodology. Each was to become the provenance of specially trained researchers who were expected to tend their "field" and leave other fields to their own specially trained scholars. A new way of organizing the academic enterprise was born.

First at the new universities, and with time at the colleges that hired the universities' PhDs, faculty came to see themselves, and were seen by colleagues, as disciplined investigators of a distinct field. No longer would a faculty member roam over whatever range of topics might take his or her intellectual fancy. On the contrary, the disciplinary scholar derived his or her insight and authority by specializing in one field. This disciplined focus distinguished the professional from the wide-ranging amateur or dilettante, who came to be looked down upon by members of the disciplinary guild.

Increasingly, these disciplinary specialists were mentored and trained in a handful of research universities. These universities took the responsibility of ensuring that the scholars in training covered the "field" and thoroughly imbibed the appropriate methodologies and attitudes proper to the disciplinary specialist. They took on responsibility of certifying by examination and by the requirement of intellectual "master pieces," known as dissertations, that the budding specialist was fit to join and contribute to the disciplinary guild. They licensed professionals to research and teach with the conferral of the doctor's degree (PhD) in a disciplinary field. By the end of the nineteenth century, the certification conferred by the PhD was fast becoming a requirement for faculty employment in universities and colleges across the country.

With these developments arose a self-conscious sense of being a member of a professional community that included fellow specialists cultivating the same disciplinary field. The community organized itself into associations, founded journals and established conferences to facilitate communication with each other, and developed a range of practices that allowed its members to evaluate and police the discipline. These professionals developed practices, such as the scholarly footnote and the literature review, whereby they acknowledged each other as fellow practitioners and demonstrated that they understood what the community was doing and how they saw their own contribution fitting in.

As a professional community, they also asserted the right of self-regulation and developed the means to police their own members. To this end, they developed practices, such as peer review in hiring, publishing, and the awarding of grants. The research universities responsible for training the next generation of specialists developed traditions and policies for mentoring and educating junior members. At each college and university, resident specialists adapted to local conditions standards regarding hiring and promotion that were advocated by national disciplinary associations. By such means, the disciplinary community of practice constituted itself and crafted ways to identify itself as a distinct disciplinary community and to maintain its peculiar goods, standards, and practices.

A key conviction accompanied and justified the development of specialized fields. Scholars came to believe that through a community's collective efforts, including peer review and other means of monitoring and evaluating individual work, a winnowing occurred that separated the better scholarship from the worse, the good from the bad, even perhaps the correct or "true" from the incorrect or "false." Under the impress of Pragmatism, some came to argue that this process of self-evaluation and correction allowed the community's understanding of its subject matter to converge over time toward the "truth of matters." This was always a provisional truth, a revisable truth that might have to give way to a new understanding as new evidence or better explanations arose out of the work of the community.

These new, professional disciplinary communities dramatically complicated a faculty member's relationship with a college (now also university). They also put an end to the broad consensus secured by natural theology that had organized and informed antebellum education. As disciplinary professionals, faculty were now responsible for the (evolving) standards and practices of their own disciplines rather than to the harmonizing assumptions of natural theology that had held the curriculum together in decades past.

Specialization and professionalization had led to differentiation and departmentalization. Much as the new industrialization and urbanization of the Progressive era created a divide between home life and work life, so too did academic professionalism suggest different concerns and ways of thinking for private life and disciplinary scholarship. Faculty now drew their self-identity and the ways they construed self and world from at least three, sometimes overlapping but still distinct, communities: disciplinary community, college or university, and church (or, in a few cases in this early period, synagogue). In the tug and pull among these three communities, the identity supplied by the church became more and more marginal to the faculty's academic persona. Well into the twentieth century, most faculty were at least conventionally religious, but their religious commitments were progressively unlikely to appear explicitly in either their scholarship or their teaching. In fact, faculty in many disciplines would often be at pains to deny that their religious conviction had any proper role in scholarship or teaching. To mention such things would simply violate disciplinary standards.

The discipline's goods, standards, and practices often differed from discipline to discipline and even within disciplines. But there were some broad commonalities, at least among the natural and social sciences and even some of the more "scientific" humanities, such as philology. Scholars in these fields championed what they termed the "scientific method," although what they meant by this has been hard for historians to pin down.[21] Most advocates of a scientific method agreed, as Roberts and Turner explain, "that the key to doing science—to producing knowledge rather than speculation—was to *think small*: to ask questions for which there were determinate and publicly verifiable answers" (emphasis in the original).[22] They chose on principle to limit their research to phenomena that were accessible to "objective" examination and verification. They not only still spent much time gathering and classifying facts, as their Baconian predecessors had done, but they also came to champion the need for hypotheses or theories (models), something their Baconian predecessors had rejected and scorned. They came to embrace the approach that historians and philosophers label "methodological naturalism," namely, the assumption that scientifically adequate explanations should be supplied by causes and factors that do not refer to the divine.

What counted within the professional discipline as an acceptable explanation had, in effect, changed. "From the vantage point of those calling for a more rigorously naturalistic methodology, the affirmation of supernatural activity in the face of mystery seemed to short-circuit scientific inquiry and to exhibit an odious form of sloth, a sin especially repugnant to good Victorians." Roberts and Turner explain, "The appropriate response to the inability to account for natural phenomena naturalistically was to solicit further scientific inquiry, not posit the supernatural. Increasingly after 1870, scientists preferred confessions of ignorance to invocations of supernaturalism."[23]

We have seen how theistic assumptions regarding God and nature, and Christian assumptions regarding nature and Scripture, tied together the curriculum of antebellum American colleges. The development of disciplinary specialization during the Progressive era led to a gradual exclusion of all nondisciplinary methods or perspectives from the work within the field. Research became "departmentalized." This meant that specialists, including those who were personally religious, came increasingly to believe that religious perspectives or considerations did not belong in disciplinary work. They were not part of the discipline's appropriate subject matter or method, however much they might infuse the scholar's private life or give meaning and background to the work he did as a disciplinary scholar (i.e., give a sense of vocation to his disciplinary work). As fields took form and divided up the expanse of human knowledge, religion was given its own acres but was no longer—except perhaps in grand metaphysical terms—seen as the universal cynosure that made sense of all. That role was now taken, more likely than not, by science, or at least the "scientific method" and the ideal of "objective" naturalistic scholarship.

## Disciplinary Assumptions

Let's now skip over the intervening history and examine some key disciplinary assumptions prevalent today. As with the case of religious traditions discussed in the last chapter, academic disciplines tend to operate with low-level assumptions that incline the individual scholar when thinking about the cosmos, human being, human obligation, and the reading of texts. Philosophers have explored these assumptions in detail, and most scholarly professionals recognize what they are, even if they may not be able or willing to elaborate on all their implications and possible difficulties. Here are a few dominant assumptions. They are chosen both for their central importance and because they line up with characteristically religious assumptions we surveyed earlier.

We begin with common assumptions about the cosmos and human being. Briefly, in much of the natural and social sciences, scholars are inclined to work within what is sometimes termed the "naturalistic assumption," namely, that "ultimately nothing resists explanation by the methods of the natural sciences."[24] This assumption applies both to the cosmos and to human beings and human social behavior. Until recently, one ideal within the natural sciences was an explanation that resembled the Euclidean model of axioms and formal deductions. This approach was abstract, universal, and formally certain. To the degree that any particular explanation fell short of this abstract, universal, and certain ideal, it was (and in some circles still is) to that degree "soft" and even "unscientific" and a lesser species of knowing and explanation. This ideal has potentially drastic implications for the social sciences, since it suggests that human conduct and human social institutions should also be explicable in terms of deterministic (or, an unfortunate compromise from some perspectives, probabilistic) cause-and-effect relationships.

Yet most scientists may not understand the "methods of the natural sciences" exactly in this abstract, universal, and formal way. More commonly, perhaps, they understand the methods of the natural science to embrace local models or theories, the careful framing of hypotheses, and rigorous empiricism. A philosopher may detect metaphysical assumptions lurking in the day-to-day practices of working scientists, but the scientists themselves may not be knowingly advancing metaphysical claims. And in the community of natural scientists, religious claims are generally out of place or are considered, as it were, extracurricular and personal, not as part of the community's approved discourse.

When we turn to the humanities and the more humanistically oriented social sciences, the situation becomes even more complicated. Few humanists or humanistically oriented social scientists feel that the scientific method, however understood, is appropriately applied to their disciplinary subject matter. Some will respect a scientific approach in the right context (say physics or chemistry) whereas others may look with skepticism on science generally.[25] Beyond this rather obvious point, it is extremely difficult to characterize the range of assumptions about the cosmos and human being one can find in the humanities and the more humanistic social sciences. They range across multiple spectra, from Romantic affirmations to skeptical deconstructions, from intensely moral asseverations to nihilistic doubts, from highly concrete narratives to broadly theoretical assertions. In the wild free-for-all that is the humanities and the humanistic social sciences, overtly religious assumptions may not be at home but they are nevertheless not automatically disqualified as self-evidently beyond the pale.

The proper role of contingent, situated narrative in naturalistic accounts has been a matter of controversy for the proponents of the naturalistic assumption and continues to be so in some circles. Yet, narrative plays a role even in the "hard" sciences of evolutionary biology and geology, and it is employed extensively in the social sciences and the humanities. We'll have opportunity to explore this and some implications later.

We find a similar variety when we turn to assumptions about morality, human obligation, and values. Across the disciplinary spectrum, scholars are of many minds on the matter of values, much less obligations. Nevertheless, even those who are squeamish about allowing moral questions into disciplinary scholarship will admit that disciplinary professional formation develops in the disciplinary professional a strong set of values regarding appropriate scholarship. Among those values can be the value of "value neutrality," or "objectivity," or "detachment" in the doing and the interpreting of research. I place quotes around these terms because they mean, and have meant, different things to scholars over the last 150 years and are still capable of arousing debate and disagreement. In science, an assertion of objectivity and even value neutrality may gain a respectful hearing from other scientists. In the humanities and some social sciences, a similar claim is more likely to garner the scholar a skeptical snort and an incredulous look that asks whether the scholar could possibly be serious.

When we turn to moral questions, conceptual problems surrounding "value neutrality" or "objectivity" escalate to a new level. Although on most campuses the range of positions is large, we continue to find some scholars and some disciplines that remain profoundly uneasy with the whole matter and wish to claim that moral questions are beyond their purview. If nothing else, moral questions, they suggest, do not and cannot meet the naturalistic assumption under which they as scholars and scientists work. Within some forms of the naturalistic framework, questions of value and moral obligation are construed as simply human constructs and expressions of subjective opinion and, hence, beyond scientific or scholarly scrutiny. Yet, when we turn to the humanities and some of the social sciences, we may be as likely to find forthright advocacy and strong moral posturing. The academy is of many minds on such issues.

Finally, in academe, we commonly read texts with the goal of understanding as part of some external purpose we have. This applies even to texts deemed sacred within one religious tradition or another. We adopt a technical orientation to the text by acquiring the full apparatus for situating and understanding the text in its context(s) and temporal and social situation. And we maintain a certain psychological and academic distance from the text, even as we may seek to put ourselves imaginatively in the place of the readers for whom it was intended. We read largely as active agents who extract insights and information from a passive text.[26] (We are speaking here of an academic ideal of a certain contested sort—the text often works its mysterious influence on us despite our attempts at academic distance.[27])

## Professional Formation into a Disciplinary Community

The educational sociologists Jean Lave and Etienne Wenger have studied in various settings the socialization process that over time turns apprentices into masters.[28] Learning, they insist, is a process in which "agent, activity, and the world mutually constitute each other."[29]

> Notions like those of "intrinsic rewards" in empirical studies of apprenticeship focus quite narrowly on task knowledge and skill as the activities to be learned. Such knowledge is of course important; but a deeper sense of the value of participation to the community and the learner lies in *becoming* part of the community. . . . Moving toward full participation in practice involves not just a greater commitment of time, intensified effort, more and broader responsibilities within the community, and more difficult and risky tasks, but, more significantly, an increasing sense of identity as a master practitioner (emphasis is in the original).[30]

Participation in communities of practice begins as legitimately peripheral but increases in engagement and complexity as the apprentice moves toward mastery. In this process, the apprentice finds motivation in becoming a member of the community of practice, in identifying himself or herself as a master practitioner.

The typical stages in academic professional formation may look something like this.[31] Our apprenticeship often begins in the undergraduate years, when we choose a major, take courses within the major, begin entertaining the thought that we might want to become professors ourselves, and are, perhaps, encouraged by our professors to continue on to graduate school. We learn by observing our faculty role models and by doing research tailored to our novice status. Our self-identity is shaped in the process. We describe ourselves to others and to ourselves as a such-and-such major, and we imitate in small, often controlled ways the behaviors and convictions of a full practitioner. We find satisfaction as we begin to exhibit the skills and values that constitute the disciplinary practice. Increasingly our continuing progress within the field requires not only intellectual engagement but also emotional commitment; we come to care about what we're learning and find that we learn best that which we care about. This process continues and intensifies during our graduate school years.

During our apprenticeship, we accumulate a wealth of experience discriminating between good and bad academic practice. We acquire this experience vicariously through reading our discipline's literature and learning its paradigmatic stories. We acquire it observationally by watching master practitioners, namely, our mentors, teachers, and senior colleagues. We also acquire experience firsthand by doing research and teaching and by receiving feedback from senior practitioners.

Gradually we are acculturated into the intellectual mind-set of a disciplinary professional. We begin with rules and maxims but soon progress to a more intuitive sense of what a particular intellectual or research situation requires. With experience, we move from a stage where we have to painfully reason through how to interpret or explain a phenomenon to a stage where we immediately see the alternatives and know the strengths and weaknesses of different approaches. With time, we acquire what the philosopher Hubert Dreyfus calls "situational discrimination" and a repertoire of responses appropriate to each subtly different situation. As we move into mastery, we come to see and know and also do with a style that is our own, a style that we can self-consciously situate within the tradition of our discipline.

In the earlier stages we can be relatively detached. But at some critical point, further progress depends on emotional engagement. To reinforce the lessons we're learning, we must feel the risk associated with our analysis of the situation and with our explanatory or interpretive choices. We need to feel elation (or at least satisfaction) when we do well, and shame (or at least discouragement and frustration) when we do poorly.

We need the guidance of a mentor, or ideally, mentors, who can help us to see the salient aspects that allow for situational discrimination, who can aid us to build our repertoire of responses appropriate to each situation, and who can praise us when we do well and criticize us when we fall short. The emotional valences of this mentor–apprentice relationship may often be unhealthy but can also play a useful (albeit sometimes perverse) pedagogical role. The more we care, the more we try to impress, the more the lessons, both positive and negative, tend to stick. We human beings are wired for learning through this kind of interaction, which starts in childhood and continues into the apprentice–master relationship and beyond.[32]

We are given relatively controlled opportunities to try out the role for which we are preparing. We do research and teach, first under the tutelage of senior colleagues, who can suggest, reinforce, and inhibit as the case requires. Increasingly we are allowed to go solo. We become, at least figuratively, journeymen and journeywomen through stints as postdocs or teaching assistants. We continue our (supervised) development through a probationary period as instructors and assistant professors. We learn by modeling our professional behavior on the examples offered by our mentors and other senior faculty. We choose among the intellectual, research, and pedagogical styles offered, favoring some and reacting against others. We gradually fashion our own style as an academic professional.

Even as we achieve tenure and enter the associate professor ranks, we still have ahead of us our masterpiece, another book or more articles that, in the opinion of our superiors, show that we are qualified to enter the rank of full professor. And even after obtaining the rank of professor, we continue to compete with others for grant funding, status, and choice of desired institution at which to do research and teach. In a real sense, our professional formation never ends, both because of the context in which we do our work and because of the standards and expectations that we have so deeply internalized over the course of our scholarly careers.

As we move through the stages from apprenticeship to mastery, we increasingly self-identify with our discipline. We think of ourselves—and others see us and treat us—as historians or chemists or sociologists, and we internalize the goods, standards, practices, and schemes that our discipline offers us. Even as we seek to move our discipline forward through new approaches, new discoveries, and new interpretations or explanations, we internalize the traditions that have been handed down to us and use them as the springboard for, and the measure of, our own accomplishments.

We also gradually internalize our discipline's assumptions on what constitutes proper knowledge, inquiry, explanation, and interpretation. In learning to exercise ever more sophisticated situational discrimination, we make second nature our discipline's take on that aspect of the world that is its subject matter. As we acquire a repertoire of interpretive or explanatory strategies appropriate to each nuanced situation, we come to look at our subject in a way that opens up some forms of knowing but may close down others. We acquire an *orientation* that shapes how we see, understand, and experience our subject, an orientation that also influences the range of our imagination and sympathy. The extent to which this disciplinary orientation spills over into how we see and understand the world generally varies a good deal from discipline to discipline, and practitioner to practitioner.

Practices within our disciplinary community are acquired in stages over a lengthy period of time and often come to influence how we see, understand, and experience aspects of the world. Let's consider briefly two examples: first, how scientists learn the practice of doing a scientific experiment and second, how academics generally acquire the practice of what might be termed the academic reading of texts.

In learning how to do a scientific experiment, we often begin in high school or even earlier, observing our teachers as they do an experiment in front of the class. At some point, after instruction in the scientific method and careful directions, we begin doing supervised experiments ourselves. Initially, the experimental steps are rather formulaic, and we are expected to get the "correct" results and are graded accordingly. This artificial preapprenticeship model often extends well into college, although an occasional science fair experiment may afford opportunities to experience the uncertainty, curiosity, adventure, and excitement of actual experiments where the answer is not already known. But if we continue in a field like chemistry or experimental psychology, we eventually get to assist a master practitioner in doing actual experiments, learning by observing, participating, and doing ourselves. With time we, too, gain Dreyfus's situational discrimination that allows us almost to intuit what results may mean, when the instruments may need realignment, why some things work and others don't, and what to do when all else has failed. From working in a lab with others, we gradually earn the right to increasing independence, eventually gaining a lab of our own. We become masters of the practice of the experiment within our discipline.

In the course of this development—which often extends over years of doctoral and postdoctoral experience—we acquire a certain way of seeing and experiencing those aspects of the world that we are studying. We gain the eyes (and the disposition, language, culture, intuitions, imagination, and sympathies) of a master practitioner. This is reflected in how we do experiments, in how we write them up, in how we view the work of others, and in how we tackle problems even outside our field.

Further, and importantly, we come to see, understand, and experience aspects of the world in a certain way. To offer a simple (or really, simplistic—but bear with me) example, consider how

1. a common experimental method—take the thing being studied apart and study one part while keeping the other parts constant ("methodological reductionism")—

interacts with

2. a way of explaining—the whole is explained by the interaction of its parts ("epistemological reductionism")—

that, in turn, interacts with

3. a view of the world—a claim that all entities in nature are reducible to their parts ("ontological reductionism"). This view may assert either that wholes are *nothing but* their parts, or, in a weaker and more commonly held form, that wholes are *determined*, causally and ontologically, by their parts.[33]

The practice of the science experiment often entails an orientation, that is, a way of understanding and experiencing the world. As with this example, the more profound practices, whether in religion or scholarship, often have epistemological consequences and perhaps even metaphysical implications. In later discussions, we shall explore how often we or other practitioners actually draw such far-reaching metaphysical conclusions from methodological (or epistemological) practices. We may more easily switch mental frameworks from context to context than the progress from methodology to epistemology and on to metaphysics may suggest. More on this possibility later.

We also begin in childhood acquiring the practice of academic reading.[34] Our formation begins even as we learn to read. It begins with observing and listening to others read, moves to instruction on how to read ourselves, involves supervised performance, and is marked by tests and rites of passage (from one reading group to the next, in elementary school; from reading book reviews to doing book reviews in graduate school; from reading books and articles to writing books and articles as professional academics). As our formation progresses, we reach a point where we have to care about what we read; we must venture emotional commitment. We must also venture our own readings (and thus risk being charged with misreadings), all to show that we can read accurately and well. We are increasingly called to put our readings into print through book reviews, articles, books, evaluations of colleagues, and so on.

Academic reading emphasizes the rapid acquisition of content useful to the purposes we, the reader, set. Most of the time, we read to acquire content useful to our research or teaching. At stages in acquiring this crucial practice, we are tested on how well we do. By the end of high school, there is the SAT. To enter graduate school, there is the GRE. Before beginning our dissertation, we must pass our general (or oral) examinations, which commonly stress our knowledge, gained by academic reading, of the major issues and literature in our field. In our specialty, we become acute reading technicians, equipped as best we can manage with a thorough understanding of the context of our text, the physically and temporally situated meaning of its words and concepts, and the critical apparatus necessary to understand and use the text for our purposes.

In comparison with religious reading, where the sacred text has its own independence, authority, and active claim on the reader, in academic reading, the text is largely a passive source from which the active reader gleans that which suits her own purposes, which are not necessarily the purposes of the text itself. We academic readers *respect* many of the texts we read, and we build grand edifices to house them (libraries), but rarely do we treat them with *reverence*. (Or if we do, someone is sure to raise the religious analogy whether in an ironic or appreciative vein.)

If you harbor any doubt that we are professionally formed into the practice of academic reading in a discipline, consider how much energy we professors spend on teaching our students how to do it and testing how well they have learned to do it. Consider how much energy is expended offering "readings" and correcting "misreadings" of each other's work. Some scholars owe much of their professional prominence to their ability to do a good (sometimes witty, frequently biting) book review. Academic reading lies near the heart of our professional competence.

In sum, the acquisition of practices stretches over considerable time, comes with deep emotional valences, and becomes intertwined with our self-identity and how others see us. As a result, we find the orientation that a practice instills difficult to question and are apt to take it as self-evident. This orientation may also entail background or control beliefs that influence what we see, believe, and accept within our disciplinary work and beyond.[35] We'll explore these implications in several later chapters, but first we need to look at the institutions that provide the most common home for disciplinary professionals—the college or university.

# Chapter 5

# Institutional Settings

Academic formation begins in institutional settings, at a doctoral (or other terminal degree, e.g., Masters of Fine Arts) granting institution. It continues in employing institutions, commonly colleges or universities but sometimes research institutes or foundations (e.g., the Institute for Advanced Studies or the Heritage Foundation) or public or private centers of research and practice (e.g., hospitals, pharmaceutical companies, and engineering firms).[1] Ongoing religious formation may also take institutional form at graduate or employing institutions that are church related (or, e.g., in the case of Brandeis University, what might be termed "synagogue- or temple related"). To grasp better the variety and complexity of academic and religious formation within institutional contexts, we need to recognize, first, that communities of practice necessarily take institutional form, and second, that in the case of colleges and universities, these institutional forms may simultaneously embody an overarching community of practice that may in important ways be distinct from the disciplinary communities of practice that are its most obvious constitutive parts.

## Institutionalized Structures and Processes

In general, we need to distinguish between practices and the institutions that embody and sustain them, even as we recognize that practices are shaped by their institutional expressions. In other words, we need both to distinguish between academic disciplines and the colleges and universities that embody and sustain them and to recognize the ways in which their institutional expression influences disciplinary practice. The two are inextricably entangled in practice but are distinct in concept.

To sharpen our vision, it may be useful to distinguish between goods that are internal and external to a practice.[2] Goods internal to a practice may be said to be uniquely defined by the practice and can only be realized by participating in the practice according to goods and standards set by the practice. To do "good history," one needs to follow the standards of excellence established and maintained by the community of professional historians. Goods external to a practice, in contrast, are not defined by the practice and can be achieved in ways that may have nothing to do with the practice. The preeminent external goods include money, status, prestige, and power. Many such goods are by nature contingent, are possessions, and follow the rules of a zero-sum game, that is, more of these goods for one person means less for another.

Institutions run on external goods. They acquire money, pay salaries, and distribute power, status, and prestige. This is not a criticism of institutions. Without concern for such external goods, these institutions could not sustain themselves nor could they sustain the practices of which they are the bearers.[3] So the entanglement of practices and institutions is not only unavoidable but also inherently problematic. Academic disciplines obviously depend on colleges and universities for material support. Perhaps less obviously, they also depend, for example, on the institutional structures of colleges and universities to maintain academic freedom.[4] But the academic disciplines must also ward against the possibility that institutional rewards—the external goods of status, money, and power—will supplant the internal goods that define the excellences of the academic discipline. Nowadays, that corrupting influence extends to government and business that would gladly turn knowledge and its acquisition to ends arguably alien to the spirit of the university itself. Academic virtues—including the self-denying, honest pursuit of knowledge—must also ward the practitioner from the corrupting influence of such external goods.

Institutional forms and institutionalized processes can directly shape internal goods. Boundaries between disciplines, for example, are maintained for more than just intellectual reasons. Departments also function as administrative units that have budgets, are authorized to hire and promote, and follow the bureaucratic logic of growth and turf protection. We academics kid ourselves if we characterize a "core curriculum" as solely an intellectual and educational matter; it is also a job protection and work allocation plan. Whatever individual faculty may think, departments naturally seek to maintain, or better to increase, "headcount."

Departments are obviously not the only institutional structure that can shape local disciplinary communities of practice. Consider the institutionalized processes and structures that guide the hiring, promoting, and tenuring of faculty. These structures and processes include the ways in which administrations give departments permission to hire, the institutional and individual negotiations involved in crafting a job description, the process by which candidates are identified, interviewed, and hired. Once hired, a junior faculty member passes through an institutionalized probationary period during which time she will likely be subject to prescribed class visits by senior colleagues who will also scrutinize her publications and her class evaluations. To be tenured, she will have to secure the support of her departmental tenure and promotion committee and pass muster with the school's tenure and promotion committee. These gatekeepers will apply not only disciplinary but also local standards in evaluating her as a teacher, researcher, and community member.

With a bit of reflection, we can easily add to the list of institutional structures and processes that not only embody but also constrain disciplinary communities of practice. Some may deal more with goods internal to a particular practice, others more with goods external to a practice. All, however, contribute in varying degrees to the formation of a master practitioner within the disciplinary community as well as within the local academic community.

## Church-Related Colleges or Universities

A few words should be said about those colleges and universities that are or remain "church related." Since the 1960s, there has been a remarkable change in the relationship between colleges and their sponsoring denominations, whether mainline[5] Protestant or Roman Catholic. Some Protestant colleges changed their relationship at an earlier date, some Catholic colleges later, but the parallels between the two groups are striking—the adoption of independent lay boards of trustees, the control of governance by the college itself, increased dependence on federal and state funding, dilution of the traditional mix of the student body, a failed attempt to use theology to hold the disciplines together, and a split between faith and knowledge. These developments have generated, in turn, recurring conferences and essays on the "identity question," hand-wringing and angry criticism, expressions of nostalgia for a bygone ethnic and denominational subculture, and so on.[6]

At one extreme are schools owned or at least controlled by specific denominations where the faculty (and often the student body) are expected to subscribe to an institutional statement of faith and belong to the sponsoring denomination. At the other extreme (at least on the church-relatedness dimension) are schools whose current tie to a denomination is strictly historic and rarely mentioned (except, perhaps, when soliciting gifts from elderly alumni).

One characteristic shared by most actively church-related institutions of higher education—as opposed to those institutions whose church-relatedness is now purely historic—is an ongoing argument about criteria for church-relatedness. Various critics have sought to prescribe what a true church-related college *is* (they really mean, *should be*, since their description is intended to be prescriptive or normative). While the "marks" vary from denomination to denomination, and from college to college, these prescriptive characteristics are usefully summarized by Merrimon Cuninggim, former executive officer of the Danforth Foundation, later president of Salem College in North Carolina, and a long-time commentator on church-related higher education. According to Cuninggim, they include

1. founding and historic association with a church denomination,
2. interrelated structure and governance,
3. financial and other support from the church,
4. the denominational credentials of the college leaders,
5. the denominational makeup of the student body,
6. the course of study (by mid-century or slightly afterward theology is supposed to play an integrating role within the liberal arts curriculum for both mainline Protestant and Catholic church-related colleges),
7. the morals and parietal regulation of campus life,
8. the provision of chapel services and opportunities for exercising moral conscience, and
9. a general religious or service "ethos."[7]

Cuninggim argues further, with some plausibility, that the insistence on such "marks" boils down to an issue of control: who is in charge of the college, its curriculum, and its campus life? The college itself or its sponsoring denomination? In most cases by the 1990s, the answer was that the college was in control, but the critics would like to reverse this development or, at the very least, deny the offending colleges the right to call themselves church-related, Christian, or Catholic.

Whatever position one ultimately takes on the criteria of church-relatedness—and there are strenuous critics of the critics[8]—church-related colleges and universities face more challenges to traditional church-relatedness than they once did. Take, for example, what Cuninggim sees as the nub of the dispute, namely, institutional control. College or university governance is necessarily more complex than it once was, and that complexity forces the institution to answer to more masters than was once the case. Nowadays, various outside institutions are in a position to exert far more intrusive control over the operations of church-related colleges and universities than founding denominations can muster. Most colleges and universities, for example, accept federal and state grant funds brought by its students and are accordingly subject to a wide range of regulations set by the state and federal governments (including the Department of Education and, more recently, the Internal Revenue Service). They also compete for foundation and other grants that often have strings attached with regard to, for example, "nondiscrimination." Colleges and universities also have to answer to accrediting bodies, and these bodies can and do direct the institution and its academic offerings in a variety of significant ways. Colleges and universities are also subject to a number of professional accrediting associations for such things as social work programs, nursing, education, and so on. Most colleges and universities have also agreed to abide by most of the rules of the American Association of University Professors (AAUP) in matters such as tenure. This list could easily be expanded, and each of the specific "marks" in Cuninggim's list could be similarly glossed.

## Colleges or Universities as Communities of Practice

Colleges and universities are not just institutions. They are also overarching communities of practice that encompass a variety of disciplinary communities. As such they can also be understood as a community of scholars that seeks to acquire, extend, evaluate, and disseminate systematic knowledge about a whole range of subjects. To that end, the disciplinary communities comprising the college or university establish and maintain certain standards of excellence that practitioners aspire to meet in pursuit not only of the various disciplines' goals but also of the encompassing community's goals. For example, while the disciplinary community of, say, chemistry may establish and maintain standards of excellence regarding research, teaching, and community service within the field of chemistry, the college or university with a particular chemistry department may itself establish its own standards of excellence regarding research, pedagogy, and community service that it broadly applies to all disciplines within its local community of disciplinary communities including chemistry.

The standards of a disciplinary community may not only closely align with the local academic community standards—this would commonly be the case with institutions classified as "Doctoral/Research University-Extensive" according to the Carnegie classification—but may also differ, at least in emphasis or weighting, in, say, liberal arts institutions that may put more emphasis on good teaching than on research and that may stress service to the local academic community over service to the larger profession. This difference is often expressed institutionally in the goals and standards applied by local tenure and promotion committees. In making this distinction, I do not mean to suggest that disciplinary communities do not recognize the importance of good teaching and local community service—obviously they do—but I do wish to call attention to the real differences in emphasis that sometimes exist between the goals and standards of a disciplinary community and a local academic community. Every scholar who goes from graduate education at a research university to teaching at a liberal arts college knows of what I speak.

When I was president of St. Olaf College, faculty, administrative colleagues, and I thought long and hard about how best to socialize new faculty into our academic community, and we deployed significant institutional resources to facilitate this transition. We tried, for example, a semester-long new faculty seminar that engaged our new colleagues in readings and discussion that dealt with everything from crafting syllabi appropriate for undergraduate education—many freshly minted PhDs are still so close to their graduate education that they tend to overestimate both how interested undergraduates are likely to be in their subject matter and how much reading they can handle while juggling three other courses—to discussing official (and unofficial) standards for tenure and promotion.

On the matter of tenure, junior colleagues need to learn how to strike a viable balance between teaching and research at an institution where good teaching and community service count for more at tenure time than may be true at the graduate institution where the junior colleague has been trained. They need also to know that good teaching and appropriate community service are necessary but not sufficient for tenure. Peer-reviewed, published scholarship is still encouraged and expected, although the quantity of such research—often a shifting standard in today's competitive job market—is likely to be less than at their PhD granting institution. To complicate their situation, the balance between research and teaching is struck differently in some departments than in others at most colleges and universities, and junior colleagues need to understand and be able to meet both departmental and school-wide standards.

Finally, junior colleagues need to understand what it means for them to teach not only at a liberal arts college, but also, in my St. Olaf example, at a "liberal arts college of the church in the Lutheran tradition." At my college, there is no faith test for faculty, but junior colleagues are generally expected to be open to, or at least not summarily dismissive of, questions that arise out of religious or spiritual concerns. Beyond that, they need to be at least aware of the ongoing argument about what it means to be a "liberal arts college of the church in the Lutheran tradition." To that end, I (and my predecessors and successors) regularly challenged the community to examine its mission. Every half-dozen years or so, the community chooses a committee of distinguished faculty that holds hearings, sponsors debates, and then delivers to the larger community its collective thoughts about the college's tradition and current identity. I have always felt that in some ways the process of deliberating one's identity is more important than the resulting statement, believing that its regular occurrence helps maintain that which is being studied.

In the last three chapters, I have attempted to provide a vocabulary and some conceptual tools that will recur in subsequent chapters. We now move from models to narratives. It is time to come to grips with our own lives as academics and what role deep conviction may have played in our own formation and scholarly choices. I have drawn a broad, rather abstract outline into which we may be able to color our own lives. If you feel so inclined, I urge you to freely color outside the lines.

# Part III

# Individuals

# Chapter 6

# Narrative Identity

We human beings generally share a strong urge to "make sense" of our lives but differ significantly in what "making sense" may entail. Toward one end of the sense-making spectrum are folks, often religious or spiritual folks, who think it crucial that all of life cohere and possess an overarching significance or meaning. For such people, this overarching coherence and significance can be achieved only through loyalty to God or to "ultimate reality" or to some grand cause that benefits something like "the larger good of humanity and nature." Toward the other end of the spectrum are those who believe, perhaps with some stoicism, that we human beings may have to settle for more modest coherence, lesser significance, and loyalty to more modest causes—and often divided loyalty at that. (Some, of course, claim to reject *any* thought of coherence, meaning, purpose, or value—but I suspect that such nihilism is more posturing than actual lived belief.)

Across this rough spectrum of sense-making, we likely share the belief that for work to be worthwhile, it should provide at least some coherence to our life, be meaningful in at least mundane if not ultimate terms, and express commitment to causes that are both worthy and larger than (or truest to) ourselves. In our individualistic and expressivist world, we may be forgiven if we hope that this work meaningfully expresses aspects of our core identity.

In this chapter, we explore how we might explain to ourselves and others how we became scholars and what we have chosen to study and teach—and what this narrative account might reveal about how human beings, and human communities, construct meaningful narrative accounts with regard to careers and work. While chance and serendipity certainly play a role in our choice of careers, the story we frame to explain to ourselves and others our career choices also tells us something about the values that we see guiding our choices. These choices should reveal something of our sense of purpose, our hopes and dreams, and our evaluation of life in general. These narrated choices may reveal both how we attempt to make sense of our lives and perhaps how, in the process, we may deceive ourselves. The stories we tell about ourselves and our career choices will also strongly reflect the influence of various communities of practice, and how each community has influenced us.

Finally, as we think through how we might narrate our career choices and the contingencies that may have opened or foreclosed possibilities, we may be better prepared to recognize that at least in the *doing* of scholarship (as opposed to the *results* of scholarship), the influence of deeply held convictions is profound and inescapable. And what is true for us as scholars and teachers is also true for our students. That being so, we need to consider how we become more self-aware and self-critical in dealing with these often subterranean but powerful influences.

## Insights that Are Exclusive to the "First Person"

Before reading much further, you should consider whether you're willing to *experience* how we narrate identity before reading about it. Appendix 2 in this book on "How and Why I Became an Academic?" can form the basis for an exercise in framing individual career narratives, either solo by journaling or in a conversation with faculty colleagues who share with each other their respective career journeys. In doing this exercise, you may gain first-person insight that an objective account can only gesture at.

The philosopher John Searle makes an important but controverted distinction between first-person and third-person ontology in the philosophy of mind.[1] Put prosaically, Searle claims that there is a significant difference between possessing information about a state of affairs and experiencing that state of affairs. With respect to narratives of identity, then, I am claiming that there is a subjective quality to constructing an autobiographical narrative that makes sense of your life that differs significantly from understanding how identity narratives work (or even how a dispassionate observer would understand how your own identity narratives works). I think that you will understand better the bearing of religious and analogous commitments on academic careers and interpretive choices if you attempt to construct a narrative that more or less satisfactorily explains your own career trajectory as best as you understand it.

There are, however, risks. You should keep in mind that those who wish to proselytize use conversation about significant turning points in their target's life both as an opening gambit to engage interest—we like to talk about ourselves and are flattered when others seem interested—and as a prime means of bringing the target to conversion through the process of reshaping his or her life narrative.[2] While I am only seeking to "convert" you to a deeper understanding of the role of (perhaps only humanly constructed) purpose, meaning, and values in our self-understanding, others may see in this exercise an opportunity for more than the enrichment of understanding. From Twelve-Step groups to the conservative Christian agenda embodied in Rick Warren's bestselling *The Purpose Driven Life: What on Earth Am I Here For?*[3] and from conventional psychoanalysis to New Age encounter groups, the telling of one's story and its gradual restructuring by interventions, questions, and the proffering of alternate plots are central activities leading to conviction and conversion.

If you still want to risk the exercises—and I hardly think the risk is that great among faculty colleagues—stop reading now and come back to the rest of this chapter only after you've undergone the exercise of constructing a career narrative for yourself either by journaling or by conversing with faculty colleagues. Or (as something of a compromise between the two because what follows may "immunize" you against some subjective insights) do the exercises after reading through the rest of this chapter.

## Making Narrative Sense of Our Lives

Let me preface our consideration of how we became scholars with some observations about conversation, narrative, and identity.[4]

### Conversation and Narrative

First, personal narrative is, in the words of the anthropologist Elinor Ochs and psychologist Lisa Capps, a "central proclivity of humankind."[5] We human beings use narrative to make sense of contingent events, both large and small. This narrative occurs in conversations, both internal and external. When we become entangled in an unexpected situation or hear about such a situation in conversation, we often start with confusion. What really happened? What does it mean? Can that possibly be what happened? To begin answering these and related questions, we converse with others, trying out explanations, receiving feedback, modifying our account, creating in the process of give-and-take an account that *makes sense*. "Akin to the virtual dialogues that take place in a writer's head in the throes of drafting a story (and which later become invisible)," Ochs and Capps explain, "conversation lays bare the actual dialogic activity through which different versions of experience are aired, judged, synthesized, or eliminated. In this manner, conversational interaction realizes the essential function of personal narrative—to air, probe, and otherwise attempt to reconstruct and make sense of actual and possible life experiences."[6]

When our sense-making conversation involves others, we tell our tale and our listeners help us shape and reshape the tale until it "fits" and makes sense to us and to them. They ask questions, proffer their own take on the events, draw parallels with similar events in their experience or the experience of others, debate the pros and cons of different interpretive strategies. "In these exchanges," Ochs and Capps point out, "narrative becomes an interactional achievement and interlocutors become co-authors."[7] Through conversation we come to understand collectively. Furthermore, and as we explore in part II, the communities to which we belong or with which we interact facilitate this conversational process by offering "standard" plot outlines and overarching schemata for making sense of life. Our conversational interlocutors are commonly also our community members. As fellow bearers of community plots and schemata, they become the cocreators of the account that an individual crafts to make sense of what has happened.[8]

Tensions and tradeoffs arise in the crafting of sense-making narratives. As Ochs and Capps explain, the achievement of coherent meaning and fidelity may compromise the actual complexity and ambiguity of the event being narrated. The all-too-human drive to achieve coherent meaning in our stories can achieve a "relatively soothing resolution to bewildering events," but in so doing, it may flatten human experience "by avoiding facets of a situation that don't make sense within the prevailing storyline." And Ochs and Capps continue, "The latter proclivity provides narrators and listeners with a more intimate, 'inside' portrayal of unfolding events, yet narrators and listeners can find it unsettling to be hurtled into the middle of a situation, experiencing it as contingent, emergent, and uncertain, alongside the protagonists."[9] We human beings vary in our tolerance for uncertainty and ambiguity and so we resolve the tension or make the tradeoff in different places in the narrative. But we all tend to err on the side of coherence when dealing with our own narrative identity, to which we now turn.

## Narrative Identity

Our identity, both to ourselves and to others, often takes the form of a story. When asked, "who are you?" our first reply is usually with a name—our story's title, as it were. But if pressed for more than a name, we narrate some part of our life. Our story may be severely abbreviated: we offer up our occupation or, if a student, our major. Such a reply is at best an implicit narrative that depends on the hearer to fill out on the basis of her own understanding the likely story behind that occupation or major. But in our own sense of self and in any more extensive sharing with others of who we are, we tell the story of our life. Our story is always selective; we touch on the "plot changes," the "turning points," the central roles we play, the crucial events or revelatory experiences that, to our minds, made us who we are. However brief or extensive, we are our stories.[10]

This narrative understanding of identity has borrowed useful metaphors from the study of narrative in literature. We speak of scripts, plots, and roles, and the improvisation that draws on the "repertoire" one has seen, acquired, and rehearsed. These metaphors can help us understand how identity forms, how it adapts fluidly to changing context, and how it accommodates constraining structure while allowing for creative, free agency.

Most of our understanding of ourselves—our scripts, narratives, plots, roles, and other items in our repertoire—arises outside ourselves and are acquired by observation, mimicry, and rehearsal. We *learn* our identity as others interact with us, as we are given scripts and assigned roles. We are *socialized* into particular roles by parents, peers, communities, and institutions. In technical terminology, this socialization may be "hegemonic" and may reflect the interest of dominant individuals and institutions.[11] We *internalize* these scripts, narratives, plots, and roles by incorporating them into our self-narrative and behavioral and conceptual repertoire.

Situations come with social scripts—how to behave in a church, what you should do when given a gift, or how to tip a waiter. Institutions come with roles—the role of a student (school), of the eldest sister (family), of a confirmand (church), of a sales clerk (shopping mall), of a professor (college and university). Context determines which roles and scripts are appropriate and which are not. The role of a sales clerk is appropriate at the mall, not at a family dinner table. The role of an elder sister is played with a younger brother, not with a boss. We spit on an athletic field, not in a church. We tip a valet parking attendant, not our teacher. Human beings are deft at changing roles and scripts as the context itself changes.

Our repertoire of scripts and roles is large and can be creatively combined—mixed and matched—with improvisorial daring and skill, creating in the process new scripts and new roles. Some aspects of the role of an "older sister" may enhance one's success as a sales clerk. Similarly, a student's role can be creatively adapted to the demands of the church. How others react to such improvisation shapes whether the script or role goes into our permanent repertoire or is abandoned.

Roles and scripts fit into plots, with a past and a future and a point. They acquire history, drama, direction, and purpose. Plots can be trifling or grand, short or extended through a lifetime—the plot of a dutiful child who disobeys her mother, an athlete who scores the winning points, a student who "brownnoses" her way through school, an unfaithful husband, a party girl, a sinner, a saint, a "brilliant nerd" who makes it big. Our movies, books, TV programs, and myths offer up plots, give rise to standard social trajectories, and offer for our edification triumphant, ironic, and tragic reversals. We become adept at fitting repertorial "text" to social context.

We come to know what to expect when given certain scripts, roles, and plots. And we come to expect that what we expect is often thwarted by circumstances beyond our control—by other people, social institutions, the natural world, and luck. As psychologist Jerome Bruner aptly observes, "narrative in all its forms is a dialectic between what was expected and what came to pass. For there to be a story, something unforeseen must happen. Story is enormously sensitive to whatever challenges our conception of the canonical. It is an instrument not so much for solving problems as for finding them."[12]

As social, contextual beings, we largely come to see ourselves as others see us. We are molded by attributions and projections, by the imposition of roles that we cannot easily resist, by the internalization of expectations, attributions, plots, and scripts. We exercise a modicum of freedom and self-determination by our improvisation, our mixing and matching of pieces of our repertoire, by trying on roles in our imagination before we try them out in our social life. But we are constrained by structure even as we aspire to free agency.

## Moral Stance

Narratives of personal experience, especially narratives of identity, always reflect a moral perspective or moral stance. In other words, they are not objective or comprehensive accounts. Rather, they offer a perspective on events, and this perspective normally manifests the moral commitments of the viewer. "Rooted in community and tradition, moral stance is a disposition towards what is good or valuable and how one ought to live in the world," Ochs and Capps explain. "Human beings judge themselves and others in relation to standards of goodness: they praise, blame, or otherwise hold people morally accountable for their comportment."[13]

A biologist may arrange the story of her research around the turning points in her search for the changes that plasma cells must undergo to become cancerous. It is worthwhile and important to understand disease process as a step toward finding a cure. An English professor may relate his scholarly donnybrook with a scholar at another university in terms of the right and wrong ways to approach James Joyce. A political scientist may explain how his commitment to rational choice theory arose out of a desire to find rational solutions in a world too easily swayed by ideologues. An artist explains how his career changed when he brought out his controversial black-and-white series depicting teenage prostitutes. Mark explains how he chose to leave a prestigious research university in order to teach at his alma mater, a small church-related college. Tara's decision to go to an urban university is explained by her engagement with Democratic politics. Jacob justifies his use of human subjects by appeal to the standards set by his university. People not only explain what they did, but also why. And the why often reflects moral considerations or standards.

Although all identity narratives reflect moral stance, moral stance is pronounced and central in narratives that arise in, and are shaped by, religious or spiritual communities. For example, in various Christian traditions, identity narratives offer a story not only of happenstance but also of choice, and the choices will commonly have a moral. Vocational narratives in the Protestant tradition, to mention an example that we'll examine in more detail shortly, "encode and perpetuate moral worldviews." In the case of conventional Protestant Christian vocational narratives, the moral worldview is teleological and providential.

Conversational personal narratives may generally concern incidents when expectations have been violated, and the conversation is needed to make sense of matters and sort the wheat of meaning from the chaff of incidentals. The identity narrative in Protestant Christianity tends to select incidents that *fit* the expectations of the vocational plot outline, and even when the events violate community expectations, they frequently foreshadow or at least prepare for a subsequent (anticipated) awareness of purpose and (perhaps) providential teleology.

Finally, with regard to moral stance, the evaluation occurs with respect to local (or community) norms of goodness. The interlocutors may not only shape their narrative to make themselves look good, but they may also, where the narrative plot outline requires in—as in certain vocational narratives in the Christian tradition—shape their narrative to criticize their own behavior and thereby deepen the significance of the ultimate conversion. In such a case, the sinful behavior described acts as a sort of negative foreshadowing or at least a backdrop to what's to come.

## Imagination

We are temporal beings. We move through life acquiring scripts, learning roles, absorbing plots, and adopting a moral stance, and as we go, we develop the narrative of our own life. We acquire a past and begin to imagine our possible futures. We come intuitively (and perhaps tacitly) to understand what is appropriate and demanded as context changes. And when habit alone does not serve, we can imaginatively anticipate the likely consequences of this script or that role in this context or that situation, and choose accordingly.

The role of imagination is crucial to the formation of identity. We are able, in the relative safety of our minds, to try out scripts, see ourselves in roles, and imagine the course of different plots. We can imagine different futures, short range and long. We can contemplate likely outcomes of this script or that role or the other plot given the particular context, the other actors present, our abilities and skills. Having tested the possibilities first in our minds, discarding those that do not "work," and choosing the one with the outcome that we think most realistic and desirable (or least unrealistic and unwanted), we attempt to enact our choice, improvising as we go. The hard edges of the world and its institutions may resist and turn our course. Other actors push back with their own scripts, roles, and impelling plots. We are shaped by the intersection of multiple narratives, public and private, social and institutional and personal. Our narrative identity arises out of this intersection.

In the chapter "Disciplinary Formation," we considered how a disciplinary professional acquires and internalizes her disciplinary community's way of seeing, speaking, and thinking about its subject matter. We suggested parallels with how one is socialized into a religious (or spiritual) community's way of seeing, speaking, and thinking. In both cases—disciplinary and religious (or spiritual)—we acquire along the way local narrative conventions and plot outlines. Since self-identity is carried in no small part by narrative, the local narrative conventions and community plot outlines help share the way we see and understand self and world.

In this regard, we need to keep in mind that different life contexts may call for (or call forth) different interpretive schemes and associated narrative plots. Even "comprehensive interpretive schemes"[14] share "headroom" in most people with other interpretive schemes, and in some cases, the contending schemes may formally contradict each other. In practice, however, they may well reside more or less happily in a person's mental tool kit. Modern humans are masters of assembling and employing ideational odds and ends, what the ethicist Jeffrey Stout calls in the moral realm *bricolage*.[15] As context changes, we pick from our tool kit what seems to work best.

For example, when a scholar who is religious talks about Creation (or the beginning of the universe), it matters whether she's talking to her 8 year old, to an adult class at church or synagogue, or to her physicist colleague who researches string theory. Human beings draw on different interpretive schemes and employ different discourses in each case. Each has its own internal consistency though it may not comport fully with the others. Another example: we scholars, especially in the natural and social sciences, may deploy reductive or deterministic theories to make our subject matter manageable and explicable. But short of extending these methodological strategies to metaphysical claims—that is, to claims about how the world works in toto—we are simply employing the interpretive scheme we share with our disciplinary colleagues.

Here is another example drawn from my experience as a Christian: academic: faculty who are churchgoers seem to be comfortable with "premodern" exegesis of scripture preached Sunday after Sunday despite knowing that biblical criticism for the last 200 years has dramatically challenged such exegesis. We don't generally expect our pastor to accommodate his Christmas sermon to the scholarship on the Marian chapters in Luke, even though we know that he has read Raymond Brown's book on the nativity stories.[16] Higher criticism is arguably inappropriate in this context.

If pressed on each example of intellectual *bricolage*, we may be willing to prioritize applicable interpretive schemes and even, in cases, admit that in a certain sense one is "right" and the others "wrong." But in real life, we are rarely forced to choose and don't feel the lack or the need. We shall explore the range of this embodied flexibility—and relaxed humility—later.

With these preliminary and conceptual observations out of the way, let's now turn to the identity narratives that professional academics might tell.

## How and Why I Became an Academic

What story might we tell if asked how and why we chose our academic career, our disciplinary field, our research specialty, and our teaching areas? In the modern West, most narratives on this question would include the following:

1. **Ability and incompetence** An account of the abilities that enabled us to do well (or at least well enough) to become a scholar and a teacher within our discipline and specialty, and perhaps also a recitation of some key skills or aptitudes that we lacked, thus foreclosing alternatives.
2. **Interests and aversions** An account of interests that motivated us in our choice of an academic career and in our choice of field and specialty, and of aversions that may turned us away from alternatives.

3. **Values** An account of the values that motivated us in our choice of an academic career and in our choice of field and specialty, including, perhaps, a desire through this profession to serve others or some larger cause. A sense of duty often motivates choice, whether the duty is to God or to one's "true self," or to some other obliging power. Strong values may also preclude choices that might otherwise suit us.
4. **Opportunities and obstacles** An account of key opportunities that assisted us and significant obstacles that hindered us in our professional journey. The narrative will likely include both contingent elements—"happy accidents" that favored us and obstacles we've had to overcome—and intentional interventions for and against this career choice by parents, friends, faculty, and others.
5. **Choices** An account of the key choices we made along the way that, at least in retrospect, look like turning points in our story.
6. **Mentors and role models** An account of key people who encouraged or discouraged us, who illustrated how it could be, or alternatively, should not be done.
7. **Confirmations** An account of how events confirmed or validated our choices. The narrative may include decision to admit us by *the* graduate school we wanted to attend, our college's or university's decision to hire us out of a competitive field of applicants, a key foundation grant, a prize for the best article or book, and so on. The confirmation may also be internal, a sense of "fit," perhaps even the outcome of reflection on our sense of alignment (see the next bullet).

Finally, there is one further consideration that can (but not necessarily) add a religious or spiritual dimension to an otherwise secular narrative:

8. **Alignment** An account that aligns our story with some larger narrative or deeper purpose. We may include among our narrative's plot elements the story of how we felt "called" to our career by something larger than ourselves. Or we might tell of how we gradually discovered our "true self" or our "deep identity" and learned to live that out through our life and work. Or we might narrate how we discovered an almost mystically fine fit between our abilities, interests, and values and the requirements of the profession we have chosen.

I'll have more to say about alignment in the sense of vocation or call in a later section.

## Ambiguity, Doubt, and the Possibility of Self-Deception

In framing the story of how and why we became academics, it is important that we also acknowledge the ambiguity, uncertainty, and perhaps even the perversity within our narrative. We may be vague about the turning points in our narrative, doubtful at times about our guiding convictions, aware of mixed motives, of several minds about the trajectory that our career eventually took, and perhaps even suspicious that our memory and our human desire for meaning may be playing tricks on us.

We should recognize that it is possible, in Parker Palmer's words, "to live a life other than one's own."[17] We can live out a life that answers more to the expectations of others than expresses our own values and sense of self. We can be true to abstract norms that dictate what one ought to do and overlook what we, as an individual with strengths and weaknesses, talents and deficits, actually can or should do. The professional socialization that we have undergone as academics constricts as well as sharpens, inhibits as well as advances, and disciplines as well as frees.[18]

For example, we should be aware that the scholarly ideals that have been inculcated in us during our professional formation may in some respects be perverse or unrealizable. In his insightful *Exiles from Eden: Religion and the Academic Vocation in America*,[19] historian Mark Schwehn traces aspects of the modern academic calling back to the sociologist Max Weber and his famous 1918 Munich University address "*Wissenschaft als Beruf*" ("Scholarship as Vocation"). In this address, Weber "self-consciously transmuted a number of terms and ideas that were religious or spiritual in origin and implication"[20] and described an academic vocation fitted to a rationalized, intellectualized, and disenchanted world. The Weberian academic is a peculiarly solitary, highly self-disciplined, and self-denying worldly ascetic, a narrow specialist striving for objectivity and detachment in a task that has no rewards except the task itself. It does require "a special kind of personality," Schwehn observes of the Weberian ideal,

> to write and at the same time will that your writing will be soon superseded. To view one's "work" as part of an *endless* process of rationalization, as a brick in an edifice of knowledge whose final shape one cannot in principle begin to imagine, is just to experience the Marxist sense of alienation from one's labor in a peculiarly acute form.[21]

Few American faculty can live according to Weber's ideal type, or will want to. But aspects of the ideal persist, for example, in the common insistence at least at research universities and first-tier colleges that our true work as faculty is scholarship, not teaching or community service. Some of us may be comfortable with this; others alienated. The Weberian ideal also lingers on in the insistence in some disciplines that our work must be objective, rational, and free of all consideration of ultimate questions of meaning or value. It persists in our dedication to narrow specialization and may stimulate a vague uneasiness with any suggestion that we should devote ourselves not only to scholarship but also to the cultivation of our students' character. Again, some of us, especially in research universities, may be comfortable with this concentration on scholarship, but others, especially in liberal arts colleges, may protest. As Schwehn aptly observes, Weber's account of the academic can be seen as "alternatively ennobling and devastating."[22] We are likely to vary according to institution and temperament on how ennobling or how devastating we find the ideal.

Many of us may nevertheless still find our primary fulfillment in the pursuit of large aspects of Weber's ideal, but for others it may in fact be a sign of both sanity and good health to find fulfillment, as ethicist Gilbert Meilaender suggests, not in our academic work but rather in "personal bonds like friendship."[23]

## How Narrative Identities May Change

Narrative identities are not static. They change over a lifetime in complicated and interesting ways. For example, one's narrative identity changes as one moves from one stage of life to the next, from, say, graduate student to professor or from child to parent. Narrative identity also frequently changes when moving from one community context to another. Recall the example of the professor who tells one cosmological story to her child, another to her adult Sunday school class, and possibly a third to her colleagues in the natural sciences. Similarly, when in the context of church, she is likely to offer a different identity narrative than when in a faculty seminar. These different narratives may cohere logically, or they may only cohere because they are fragments, the *bricolage*, of an actual individual life.

Identity narratives may also change when, for whatever reason, the narrator is convinced to adopt a new plot outline. Even when the dissonance between narrative and experience becomes high, it normally takes the availability of a more plausible plot outline before the switch occurs. Rarely do we abandon even an incoherent or weakly applicable plot for no plot at all. Such anomie tends to be too painful. But when a more persuasive alternative is offered, we may suddenly renarrate our sense of self. In such large-scale renarrations, events in our life that once were significant are now treated as incidental and events that meant little to us before suddenly become hinges around which our lives are seen to have turned. One is "born again" or "sees the light" or "becomes disenchanted and alienated." Such massive reorganizations of narrative identity go under the heading "conversion," and that conversion may be to any number of religious, spiritual, or secular ways of construing self and world.

## Self-Deception

Finally but importantly, some have suggested that retrospective narratives invest happenstance with unwarranted significance.[24] According to this skeptical view of autobiographical narrative, we project meaning onto chance. We attribute significance to offhand comments or decisions. We impute a "rightness" to a particular career path that—had we been able, as it were, to replay the tape of time—would have struck us as just as "right" had it taken a different course. We invent meaning by tendentious selection, ignoring inconsistencies that do not fit our desired plot. Our life stories, so it goes, are really fictions, crafted from inherently meaningless odds and ends, a remarkable but self-deceptive product of individual and social construction.

This goes double for attempts to fit one's life into a larger, often religious story. It is at this point that literary devices may mislead. Consider, first, foreshadowing. In a personal narrative, we often make use of foreshadowing. Past events are interpreted as harbingers of things to come. Foreshadowing plays a major role in different religious traditions. For example, in the Christian New Testament, accounts in the Hebrew Bible are regularly interpreted as pointing to, that is, foreshadowing, events in Jesus' life. Foreshadowing also plays a crucial role in vocational narratives, which we'll be examining shortly. But whereas foreshadowing may be a deliberate device used by an author, when applied to live experience it depends on attributed notions of purpose, direction, and causality that may entail more deception than discernment.

Here's another device that critics think may mislead. Some theorists have developed by analogy with foreshadowing a category they term "backshadowing," in which knowledge of outcomes is used to judge past behavior on the grounds that participants should have known what was coming. So, for example, backshadowing operates when students of the Holocaust argue that Jews who remained in Germany after Hitler's rise to power should have known what was coming and are, therefore, in some way complicit in their eventual suffering.[25] While backshadowing in this example is truly perverse and, arguably, logically incoherent—what is "obviously" coming is normally only "obvious" in the rearview mirror—the practices of backshadowing and foreshadowing are, in fact, quite common in identity narratives, especially those informed by religious belief or imagination. But again, we must ask whether they work through deception or discernment?

Those who see more deception than discernment in the use of these devices have proposed a third category, what they believe is more realistic, namely, "sideshadowing," which, as Michael Bernstein puts it, "refuses the tyranny of all synthetic master-schemes."

> Instead of the global regularities that so many intellectual and spiritual movements claim to reveal, sideshadowing stresses the significance of random, haphazard and inassimilable contingencies, and instead of the power of a system to uncover an otherwise unfathomable truth, it expresses the ever-changing nature of that truth and the absence of any predictive certainties in human affairs.[26]

For such critics, meaning is purchased at the expense of authenticity, pattern and direction at the expense of actual contingency and complexity.

One of the major distinctions between religiously informed narrative and secular narratives is the degree to which narrators and the community to which they belong are comfortable with, or even believe in, coherent, purposive plots, much less "synthetic master-schemes." However one comes down in this debate, it is worth stressing that narrative depends unavoidably on selection, on the choice between, in the language of Gestalt psychology, figure and ground. At issue, then, is what is properly foreground *for a particular purpose*, and what *for that purpose* is properly irrelevant or background. If the purpose changes, the selection may well change also. In my view, it is not inauthentic to select because one cannot but select. It is inauthentic only when the selection excludes elements that are significant *for the purpose at hand*. But my view will be contested by others.

Skepticism regarding retrospective narrative, especially narratives that attribute pattern to life events, may strike you as either on target or wildly off the mark or somewhere in between. As good academics and as self-reflective human beings, we must be careful of bias, self-deception, and outright error in our identity narrative as in our scholarly work.

## One Religious Variation: The Notion of Vocation

When Weber speaks of "scholarship as vocation (*Beruf*)," he is offering a secular variant on a Christian concept—"vocation" or *Beruf*—that he inherited from Martin Luther and the Protestant tradition. The term "vocation" is derived from the Latin *vox* or "voice" and *vocare* or to call. One can speak not only of "a" vocation or call, but also of vocations or, less awkwardly, callings with an "s."

When we tell our story in terms of a "call" (or "calls," plural—lives often change directions), we are commonly offering a teleological account: our career trajectory had, we think, an end (or ends) predetermined perhaps by God or by the nature of our innermost being. The experience of such a call (or calls) may come from outside or from deep within. The call or calls may prompt us to become something other than what we once were, or to become what we feel deep down we truly are or should be.

On the other hand, when we limit our account to one of fit, we may be offering a causal but not necessarily teleological account. Our talents suit us well to what we are doing, but that's all we're claiming.

When we construct an identity narrative (e.g., about how we became academics), we should remember that the vocational plotline that may nicely organize our own story may be inappropriate for others, even outrageously so. Finding a sense of call in one's occupation is likely to be far easier in today's society for middle-class academics than for folks confined to the lowest rungs of the economic and class hierarchy of our society. Just making a living may be hard enough without requiring that it be in some larger sense meaningful or fitting into a coherent narrative. But even in such cases, a sense of calling is possible but more likely in areas such as family—a sense of meaning in purpose in marriage or in raising a family or both—or community life.

Whether we offer a religiously teleological account or a secular narrative of "fit," the tales we tell of our career usually reflect what we value, what we find personally meaningful, and what we hope and plan to accomplish with our lives—along with an admixture of happenstance and chance. The plot variations in such narratives are many, but they all entail making sense of our lives, and making sense in this context inevitably raises issues of purpose and value. Whether we understand how and why we became academics in spiritual or secular terms, the subjective elements of our account—questions of value, meaning, and purpose—lie also at the heart of religious or spiritual narrative. In other words, we are operating in a domain where religious (or spiritual) traditions have commonly exerted considerable influence and arguably continue to offer significant insight.

Having raised the issue of vocational narrative in a religious (or spiritual) key, I am now going to offer some history and commentary. In the following, I seek to indicate, first, that even in its traditional Protestant form, vocation referred just as well to family responsibilities and the responsibilities of citizenship as to responsibilities within an occupation, and, second, that small "c" calls can and frequently do change during the course of a lifetime. Only rarely are calls unambiguous and singular within the secular realm, even from a religious perspective.

## Excursus on Vocation

In the secular version of this vocational plot, our talents and interests have fitted us for this profession, our commitments and values have guided our choices leading to this profession, and "happy accidents" have facilitated (and "unfortunate obstacles" hindered) us along our way. By contrast, in the familiar religious variant of this vocational plot, either God or some "higher power" has "called" us to this profession, either from without or from within. The call may be complex and discernable only in retrospect, but it generally includes several elements. God grants us skills or abilities that suit us for the profession. God grants us interests that incline us in the "right" direction. God appeals to our better nature by offering in the profession, rightly lived, clear opportunities to show forth our love of God and fulfill our obligation to serve our neighbor. Finally, God guides us by opening up some opportunities and shutting down others. God may even use our weaknesses or selfish inclinations to bring about God's ultimate purpose for us.

Sometimes, of course, we do not feel the call at all, but discern it only when looking back on our lives; a pattern emerges only as we scan our story. The great theologian and bishop of the fourth and fifth centuries, Augustine of Hippo, wrote his *Confessions* as a tale of the way in which God called him to Christian service using even Augustine's sinful desires and the malice of others to shepherd Augustine toward the destination God intended for him. Only in retrospect did he see, for example, how his desire for fame and success as a rhetorician led him to one of the greatest rhetoricians of that time, Ambrose of Milan, who, providentially, was also the bishop of Milan and perhaps the only man living who could resolve Augustine's doubts, still his questions, and return him to the Christian faith of his mother.

The Protestant version of vocation arose out of the sixteen-century Protestant Reformation and the theology of the German reformer Martin Luther. As mentioned, the notion of a "call" or "vocation" goes back into Christian antiquity. People in Martin Luther's day still spoke of a "call" or a "vocation" in traditional terms as a call to the office of priest, monk, or nun. God, it was thought, "called" people *from* a life in the world *to* the more demanding, and spiritually superior, life of the clergy. Other occupations were not "callings" or "vocations" in this special sense.

As it happened, Martin Luther's new understanding of justification by faith alone transformed this understanding of vocation or calling. In accord with his understanding of St. Paul, Luther insisted that Christians were justified by faith apart from works of the law. They were made right with God, that is, reconciled and justified, not by a process of spiritual growth accomplished by the doing of good works in a state of grace—the leading theological view of Luther's day—but solely through Christ's death on the cross. They were justified when they accepted in faith and trust God's promise that Christ has died for them. Human beings can do nothing on their own behalf. Even faith in God's promise of salvation through Christ is a gift of the Holy Spirit and no psychological work one can perform.

This new understanding of justification put the notion of vocation in a different light. As long as Christians understood justification as a process of spiritual growth, it made sense to say that the life, say, of a nun was spiritually superior to the life of a wife or a housekeeper, or the life of a priest was spiritually superior to the life of a butcher or a baker or a candlestick maker. The nun or priest was living a life that promoted greater spiritual growth than did the life of a lay person. But if, as Luther put it in Latin, salvation comes *extra nos*, that is, from outside of ourselves, then a Christian's salvation depends not on what he or she does but on what God has done for them. Everyone, be she a nun or a housewife, be he a priest or a butcher, depends equally on Christ's reconciling sacrifice.

This conviction led Luther to argue that all occupations, if done in faith and love toward the neighbor, were equal in God's sight. The housewife, or butcher, or baker is pursuing a vocation, a calling, that is equally pleasing to God so long as it is lived in faith in God's promise through Christ and in loving service to the neighbor. In short, Luther took the notion of a "calling" or "vocation" to the superior life of the clergy and recast it, insisting that before God all Christians were equal and all occupations equally pleasing to God when done by faithful, loving Christians. This Lutheran doctrine of vocation has been taken up and elaborated by other Christian groups.[27]

In Romans chapter 12, Paul writes about the variety of ways in which Christians are united even in diversity.

> For as in one body [Paul says] we have many members, and not all the members have the same function, so we, who are many, are one body in Christ, and individually we are members one of another. We have gifts that differ according to the grace given to us: prophecy, in proportion to faith; ministry, in ministering; the teacher, in teaching; the exhorter, in exhortation; the giver, in generosity; the leader, in diligence; the compassionate, in cheerfulness. (Rom 12:4–8 [NRSV])

To sum up this excursus on the Lutheran Christian understanding of vocation, Christians are all priests and are all Called (with a capital C) through their baptism to love God and serve others. The form their individual calls (with a small c and a pluralizing s) take reflects their talents, interests, and opportunities. This Pauline insistence that the body of Christ contains a variety of gifts combined with the originally Protestant notion that all licit occupations were equally pleasing to God contributed greatly to a positive evaluation of a secular culture.

# Chapter 7

# Inclinations

In this chapter, we examine how our core convictions may incline us to be favorably disposed to some interpretations or explanations and unfavorably disposed to others, *even as the public explanations we offer for our interpretations or explanations do not explicitly cite these core convictions*. We also consider how, as conscientious scholars, we strive to be aware of these influences and to compensate when they may mislead us. As with the last chapter, I propose we approach these questions through narrative, making sense of our interpretive or explanatory choices through story.

Of course there is no one pattern or narrative plot that will fit each and every one of us. Some scholars are likely to be more sharply inclined by their core convictions than others. Some scholars belong to traditions—religious, spiritual, political, or philosophic—that expect us to draw connections between our core convictions and scholarly interpretation; others do not. Some scholars will be in disciplines and specialties that offer more interpretive or explanatory leeway than do others. Natural scientists, for example, arguably have the least interpretive leeway, although they have more leeway than some will admit. By comparison, social scientists have far more leeway, in no small part because they are studying what the philosopher John Searle has termed observer-relative entities and because, now borrowing from the philosopher Ian Hacking, social scientists must deal with "looping effects" where their subjects—human beings, cultures, social institutions, and so on—can and do change in response to study results.[1] Finally, the humanities and the fine arts have the greatest leeway, largely because they are in the business of making meaning meaningful, a breathtakingly wide-open enterprise.

In what follows, I proffer some distinctions and considerations that I have found helpful in thinking through such matters for myself. I encourage those of you who see things differently to employ your own distinctions to reframe how core convictions may influence interpretive or explanatory choices. The point of the exercises is to become aware of influence, not to agree on the best model for understanding that influence.

## Interpretation and Explanation

In general scholarly usage, *interpretation* carries the sense of laying out the meaning or significance of something (or meaning*s* or significance*s*, both plural). The "something" might be the world as a whole or some aspect—a text, or an action, a period of history, a symbol, or a gesture.[2] *Explanation* carries a stronger causal connotation than does *interpretation*, of making sense by showing why or how something occurs. An *explanation* in this sense is a species of interpretation, namely, a type of interpretation that lays out the meaning or significance of something by explaining why or how that something occurs or has occurred.

Although we can rightly speak of interpretations that are causal and explanations that are not, I would urge us for the purpose of discussion to maintain this slightly artificial distinction. I do this because we may wish to distinguish among the range of academic disciplines on this point. The humanities, for example, are more interpretive in this restricted sense than are, say, the social sciences; similarly, the natural sciences concentrate more on explanation in this restricted sense than do the social sciences. Yet even under this artificial distinction, all the disciplines engage in noncausal interpretation and all leave varying room for causal explanation.[3] For this reason, and for economy of exposition, I shall occasionally employ the shorthand "interpretation" to stand for both interpretation and explanation in these artificially narrow senses.

For our purposes, this distinction between interpretation and explanation should suffice without adding extra categories such as *theory*, which in the academic world is generally a subset of *explanation* that carries the implication of coherence and, often, "testability."[4] A theory may either be "more or less verified or established explanation accounting for known facts or phenomena" or "a proposed explanation whose status is still conjectural."[5] Note the circularity of these rough definitions. Note also that with these distinctions between interpretation, explanation, and theory, I am drastically simplifying a matter that philosophers continue to debate. But again for our purposes, these rough-and-ready distinctions should suffice.

## Background Beliefs

A wide variety of beliefs operate in our mental "background." These background beliefs allow us to attend to, reason about, and evaluate matters that are, as it were, in our foreground.[6] These background beliefs are generally taken for granted, although we can always self-consciously lift them up into the foreground and put them to the question.

Among these background beliefs are beliefs that bear on how we evaluate scholarly interpretations and explanations. They demarcate, for example, what we think should qualify as "data" for our scholarship. They also delimit what we consider to be acceptable interpretations or explanations within our discipline broadly. Crucially for the purposes of this discussion, these background beliefs commonly guide our personal choices among contending interpretive strategies. They determine which interpretive strategy we find most appropriate or convincing or "true" to employ in our own scholarship and teaching.[7]

In part II, we explored how academics are socialized into disciplinary, institutional, and (some of us) religious or spiritual communities of practice. We also discussed how various of the more profound practices—we illustrated our point with brief examples of Christian prayer, the science experiment, and the religious or spiritual and academic approach to (sacred) texts—may influence how we see, understand, and experience aspects of the world. They may limit or extend our imagination, shape our intuitive processes of thought, and incline us to react positively or negatively to certain arguments or experiences. Obviously, then, these profound, socializing practices will likely play a prominent role in shaping background beliefs that bear on how we evaluate disciplinary interpretations.

In some cases, scholars have carefully thought through their interpretive choices and understand how they relate to core convictions, both disciplinary and religious (or spiritual). These scholars self-consciously choose among the contending approaches within their discipline to employ (and champion) those that fit best with their convictions regarding the nature of the cosmos, morality, and human being.[8] They can state explicitly what their core convictions are on these broad issues and how they bear on their interpretive choices.

With other scholars, the role of background beliefs may be more intuitive than obvious.[9] Scholars may not always be able to give explicit articulation to the background beliefs that shape their interpretive choices. Instead, their background beliefs, including those derived from their core convictions, may appear through an intuited sense of rightness or wrongness, of attraction or aversion, of comfort or discomfort—in short, in a sensed inclination toward one position and against another. It may take some careful introspection to get a better handle on why the inclination runs one way and not another.

In making these distinctions I am drawing on a most helpful analysis made by the Yale philosopher Nicholas Wolterstorff. In his *Reason within the Bounds of Religion*[10] Wolterstorff argues that scholars employ "control beliefs" to guide them in weighing and devising scholarly interpretations, explanations, theories, and claims. These control beliefs, according to Wolterstorff,

> include beliefs about the requisite logical or aesthetic structure of a theory, beliefs about the entities to whose existence a theory may correctly commit us, and the like. Control beliefs function in two ways. Because we hold them we are led to *reject* certain sorts of theories—some because they are inconsistent with those beliefs; others because, though consistent with our control beliefs, they do not comport well with those beliefs. On the other hand control beliefs also lead us to *devise* theories (emphases in the original).[11]

While Wolterstorff here speaks of theories, I believe that his analysis applies equally well to the broader categories of scholarly explanation and interpretation. And while Wolterstorff speaks of *control* beliefs, I prefer *background* beliefs, largely because I suspect that these beliefs may more *incline* us in certain interpretive or explanatory directions than directly *control* what interpretations we accept or reject. True, this softened distinction may resonate with some scholars, some disciplines, and some traditions more than with others. But it is the general concept that I want to establish here.

Whether we speak of *background* beliefs or *control* beliefs, Wolterstorff's analysis can assist us with our analysis of how we function as scholars (and some of us, as scholars who happen to be religious or spiritual). Wolterstorff explains that some control beliefs derive from a scholar's core convictions—in Wolterstorff's Christian focus, what he termed a Christian scholar's "authentic Christian conviction."[12] Others will be shared by secular scholars in their fields and will have been acquired in the course of their professional formation in graduate school and elsewhere. When conflict arises between these control (or background) beliefs and developments in scholarship, the scholar will frequently reject the scholarly view. But in some cases, and justifiably so, Wolterstorff argues for the Christian case, the results of scholarship may lead the scholar instead to revise what constitutes his or her core convictions.[13] Some widely held Christian beliefs, for example, have rightly been abandoned in the light of scholarship.[14]

To sum up, core convictions functioning as background beliefs commonly guide our choice of interpretive strategies to employ in our scholarship and teaching. We may be self-conscious about this guidance and have systematically thought through how our core convictions bear on our scholarship. Alternatively, the guidance may be more intuitive than explicit, more inchoate than systematic. In either case, core convictions functioning as background beliefs can influence our scholarship and our scholarship, in turn, can influence our core convictions. The interaction between core convictions and scholarship can be harmonious and helpful, productive of new insight and a sense of congruence. But it can also cause discomfort and dissonance.

## The Story of Our Interpretive or Explanatory Choices

What story might we tell if asked how our core convictions may have influenced our interpretive choices? The following is meant to be helpful, not prescriptive. Whatever works to get at the influence of core convictions on interpretive choices, use it!

### A Self-Conscious Reconciliation of Core Convictions and Scholarly Choices

Some of us may be able to provide a well-articulated account of how we self-consciously reconciled our core convictions with our interpretive strategies (and vice versa). We may tell the story of how we thought through our relevant core convictions, determined their logical implications, and asked ourselves whether a particular interpretive or explanatory strategy was logically consistent with our convictions. For such an approach to work, we had to be self-aware of our core convictions and their entailments—and not all of us are. To employ such an approach, we needed to be temperamentally of a systematic disposition with a desire for a rather thoroughgoing consistency—again, not all of us are so disposed or have such a need.[15] In this context, it is worth noting again that some religious traditions and some academic disciplines expect and encourage such self-reflection and consistency more than do others. If you've gone through this process, you're in a fine position to relate to colleagues how it was done and what it has meant for you and your scholarship.

But most of us, I would suggest, are liable to go through an analysis of core convictions and their entailments only *after* we've identified intuitively or emotionally that there was something in an interpretation that seemed to us at some profound level eminently "right" or dramatically "wrong." Our account is retrospective—and with all the dangers of self-deception that such accounts risk.[16] We come to see in the rearview mirror that certain interpretive strategies comport well or fail to comport with our core convictions. We can then tell a story of why we chose one strategy over another, or, alternatively, why we modified this core conviction or that. And even with these insights regarding a specific interpretation or within a limited domain, we may have no interest in doing a broader, systematic analysis. We believe that *this* conviction decided our choice between *these* interpretive strategies (or *this* conviction was modified because of *those* scholarly findings or considerations), but we have not thought through in any systematic way the implications of all our core convictions for our scholarship or teaching—and may well have no desire or perceived need to do so. If you're in this camp, or if you've never given much thought one way or another to the relation between core convictions and scholarly interpretations, the next section suggests a heuristic that may help surface connections.

Again, some traditions and disciplines expect the scholar to think through such connections; others do not. I am not advocating that everyone develop a systematic worldview that reconciles core convictions and scholarly approaches. That is a personal as well as an intellectual decision. But I do urge readers to do some introspection about possible connections to secure some first-person insight into this phenomenon of interpretive choice.

## A Heuristic for Surfacing the Influence of Core Convictions on Scholarly Choices

I am going to suggest an approach to discernment that relies on being mindful of how we react emotionally or intuitively to interpretive or explanatory strategies (or theories or models). Again, I shall use "interpretation" to cover all these possibilities. But first I should preface this approach with a few disclaimers.

First, some may object to a heuristic for surfacing the influence of core convictions that relies on an emotional response.[17] When the core conviction involved in an interpretive decision entails moral or ethical judgments, using emotional responses as a heuristic may suggest that the emotions are the ultimate basis of such appraisals. While I would insist that moral and ethical judgments have an emotional aspect—a claim I think difficult to deny—for the purposes of these discussions, we need only acknowledge this aspect while remaining agnostic about claims, for example, that ethical or moral judgment are essentially emotive reactions with little or no propositional content.

Second, some may prefer a heuristic that stresses intuitions rather than emotions, on the (debated[18]) assumption that intuitions can make propositional claims while emotions do not. Again, we need not enter into this multifaceted dispute. Use whatever works in your case!

Third, the following questions make no attempt to reveal all the considerations that go into our evaluation of interpretations (e.g., how we evaluate the "fit" between an interpretation and the known data that it purports to interpret). They simply assume that irrespective of the interpretation we choose, we are able to offer good and sufficient evidence (or other relevant considerations[19]) to support the interpretation itself. They assume, further, that we probably said nothing at all about why the interpretation also commended itself to us for additional, nonstated but deep-seated reasons. These questions are at best heuristics intended to help us better discern how core convictions can incline toward some interpretations and away from others.

1. **Emotional or intuitive response?** Do we ever respond emotionally or intuitively to an interpretation (or explanation) offered by another scholar or even one crafted by ourselves (which we are now evaluating as if crafted by someone else)? What is the tone of this response? Do we feel great attraction or aversion, a sense of profound "rightness" or "wrongness"? Or is our reaction more one of comfort or discomfort? I want to suggest that such reactions suggest that background beliefs may be at play, including, perhaps, some core convictions. In any case, once we have identified that we're reacting emotionally or intuitively to an interpretation, we need to probe further to investigate why we may be reacting in such a manner.

2. **Interpretive strategy or specific application?** Can we determine whether we are reacting to the interpretive strategy that informs the specific interpretation or to the specific interpretation itself? If, say, we have reacted emotionally or intuitively to a reductive explanation, are we expressing an underlying objection to reductive strategies generally or are we reacting to the specific application of reductive strategies to, say, human behavior? We may applaud reductive strategies in particle physics and object to reductive strategies in explaining human intentionality.

3. **Interpretive strategy or its implications?** If we are reacting to the interpretive strategy, can we determine whether we are reacting to the strategy itself or to its potential implications? This distinction may be a bit artificial, but consider again the case of reductive strategies. We may be philosophical holists and question all reductive strategies. More plausibly, we may applaud the effectiveness of reductive strategies in particle physics but feel uncomfortable because the strategy may seem to entail some form of atomism, bottom-up causality, and even determinism that conflict with our metaphysical commitments regarding human being (which, of course, we may only intuit; most of us are not metaphysicians with a clearly spelled out worldview).

4. **Which implications?** If we are bothered by the implications of the strategy, are we made uncomfortable by the moral implications? by the metaphysical implication? by underlying assumptions and their

implications? by a sense that explanatory levels have been confused with untoward implications? something else? Again, a few examples may make these distinctions clearer (if the examples fall flat, make up better ones for yourself from within your own discipline).

(a) **Moral implications** Discomfort with economic theories of "efficiency" may arise because "efficiency," seen within economic theory as a good to be pursued, may lead to significant unemployment and economic dislocation, a cost that some may find morally objectionable.[20]
(b) **Metaphysical implications** Discomfort with computational models of the mind may arise because they seem to entail determinism and to foreclose the possibility of free will and moral responsibility.[21]
(c) **Underlying assumptions** Discomfort with certain psychological theories because they rest on assumptions about human nature and psychological health that conflict with other notions of human nature and psychological health—for example, a belief that "self-development" is the preeminent ideal that the individual should aim for versus a conviction that human flourishing is best promoted by self-discipline, community, and the control of selfish impulses.[22]
(d) **Level of confusion and its implications** Discomfort with an attempt to explain human beliefs by reference to underlying brain states. Some scholars would see this as a confusion of explanatory levels. Although human intentions and other mental states arise from underlying brain processes and depend on the underlying states for their subjective existence, they may not be usefully explained or understood by focusing on neuronal activity. Consider an inquiry into anti-Semitic attitudes. As the philosopher Mary Midgley dryly observes, "there is no obvious reason why physical details about neurons in the brains of anti-Semites could ever be relevant to the problem."[23] If, however, the inquiry is about brain pathology that may lead to paranoia (regarding Jews and others) it may be appropriate to look for an explanation at the neuronal level. In this case (and many like it), the confusion of explanatory levels may have untoward implications for understanding human pursuits such as meaning making, valuing, and intending.

5. **Attractions and comfort** My examples have been limited to cases of aversion or discomfort or a sense of "wrongness." But sometimes we react to a proffered interpretation with deep emotional or intellectual satisfaction. In such cases we should suspect that key background beliefs, perhaps even core convictions, are also involved. Rather than running through the

list again with examples where the emotional or intuitive response is positive rather than negative, I'll leave this exercise to the reader. The point to remember is that in either case, core convictions may be at work. And in both cases, the conscientious scholar needs to be aware of these influences and be willing to compensate when they may be unhelpful or misleading.

After you have identified the source of your reaction (and there may be multiple sources), you need to ask the final questions:

1. **Core convictions?** Is there something in my core conviction that accounts for this reaction?
2. **Interpretive choice?** Did this reaction influence my interpretive or explanatory choice? Did my core convictions incline me to favor one interpretation over another? Or, alternatively, did this reaction incline me to reexamine and change the core convictions that produced it? Influence can run either way. We may reject an interpretation because it comports poorly with our core convictions or we may modify our core convictions if the interpretation and its fit with known evidence (and other disciplinary standards) strikes us as overwhelmingly persuasive. Many Jews, Christians, and Muslims have, for example, modified their understanding of the Genesis creation story in the light of the findings of modern geology and evolutionary biology.

In closing this section, I want to emphasize that we're considering how core convictions may influence interpretive or explanatory choices. We must not confuse the *interpretation* we chose with the *reasons* why we may have been inclined to choose it. In most cases, our core convictions may incline us to favor one interpretation over another, yet the cogency of our interpretation depends on its internal logic and plausibility, its fit with the material, its adherence to scholarly standards and conventions, and so on. It does *not* depend on the validity of our core convictions. In fact, only rarely in our interpretive or explanatory accounts are our core convictions even mentioned, much less invoked as warrants for the interpretation we offer. We discuss the rare exceptions in the chapter "Reticence."

## Unduly Subjective?

Some would argue that by paying attention to these emotional or intuitive reactions, we are being unduly or even inappropriately subjective. Are we in some way violating the canons of good scholarship, when we allow emotions or intuitions to influence our choice of interpretations? In some cases, perhaps; in most cases, not at all.

First, let me repeat that an interpretation or explanation rises or falls on its internal cogency, its fit with the relevant data, and its adherence to appropriate scholarly standards and conventions. We must not confuse *origins*—why we favored one interpretation over another—with *results*—the interpretation or explanation itself.

Second, emotions and intuitions play an indispensable role in all thinking. As the neurologist Antonio Damasio has illustrated with cases where injury or disease has destroyed an individual's emotional capabilities while leaving unharmed his or her higher cognitive functions, these emotionally disabled individuals are unable to exercise good judgment and become shallow, detached, and indecisive in their thinking.[24] And the philosopher (and chemist) Michael Polanyi (among others) has shown how we are able to achieve insight into wholes through an integrative grasp of particulars without being able to specify all the particulars or give discursive principles for how we intuitively related them one to another.[25]

Third, and perhaps most to the point, on important matters, an emotional or intuitive inclination is unavoidable and nearly always consequential, but need not be injurious. When we ignore the influence of emotions or intuitions in the belief that we must prescind from such reactions to be "objective" scholars, we deceive ourselves and increase the probability that unacknowledged subjective elements in our interpretive or explanatory choices will have harmful consequences. In the next section we'll examine briefly how we, as conscientious scholars, attempt to be aware of such influences and compensate when they threaten to mislead us.

## Being Aware of and Compensating for "Bias"

If Damasio and Polanyi are correct, emotions and tacit intuitions may be crucial for sound judgment among consequential issues. Yet we fear, and with reason, that emotion or intuition may cloud our eyes and prevent us from distinguishing the sound argument from the specious, or, more subtly, the more probable from the merely plausible. If we cannot do without our emotions and deep intuitions—and could not escape them if we wished, short of a cranial injury or fatal disembodiment—what can we do to minimize unhappy consequences?

Let's stipulate at the outset a scrupulous honesty regarding the material on which our scholarship is based—data, facts, evidence, texts, or whatever. The issue here is compensating for harmful bias, not scholarly fraud. Further, let us assume, as most scholars today do, that we cannot achieve some Archimedean point where we are free of all subjective taint, able to overcome all limiting perspective, possessed of a "God's-eye view" or a "view from nowhere."[26] What, then, might we do?

Each discipline has its methods for minimizing pernicious effects of personal bias. For our purposes, it may suffice to mention four general considerations:

1. **Awareness** If we are unaware of our bias, we are in a poor position to compensate for any missteps it causes. The material in this chapter is intended to make us more aware of how some of our most deeply held convictions have influenced our interpretative choices.

2. **Acknowledgment** A good first step is to alert our readers or auditors "where we're coming from." This compensating move makes more sense where interpretive or explanatory leeway is great—for example, in the humanities and much of the social sciences—and less (perhaps no) sense in fields such as physics, where interpretive or explanatory leeway is greatly circumscribed by subject matter and generally accepted method, and where in any case, emotional reactions or intuitions are likely to be less, and to be less improper (absent intentional fraud).

3. **A fair treatment of alternatives** Especially in the more interpretive disciplines it is good practice to describe the major competing interpretations and evaluate their strengths and weaknesses in relation to one's preferred interpretation. The operative word in this unavoidably approximate process is "fair." My rule of thumb, in principle, if not always realized in practice, is that those who advocate the alternative will recognize their position in my description and feel that I have described the issues accurately, even though, for whatever benighted reasons, I have failed to surrender to the compelling logic (or interpretive or explanatory persuasiveness) of their case.

4. **Community standards and policing** Finally, and we'll return to this point in more detail in the chapter "Community Warrant," we should normally attempt to meet the standards of our disciplinary community and count on colleagues, and on the disciplinary community as a collective with its institutional expressions (journals, reviews, conferences, and so on), to straighten us out when we may have been led astray by powerful convictions. As we shall see, the epistemological function of disciplinary communities is not unproblematic, but at its best it can offer another curb on idiosyncratic and misleading inclinations.

I realize that these suggestions (except, perhaps, the fourth) may seem self-evident. As professionals, we have been trained thoroughly regarding the pitfalls of bias within our particular disciplines. We regularly review our colleagues' work, and are reviewed by them in turn. We explore in this give-and-take the problems of bias and their possible remedies. (Or, alternatively, we've asserted, or have had asserted to us, the ineluctable nature of bias, especially as a function of social location, and the power dynamics that underlie calls to "objectivity.") Each of us can, with some reflection, offer an extended consideration of bias within our discipline and its significance and treatment. The importance of this issue may lie less with our own understanding and more with the understanding we awake in our students.

# Part IV

# Implications

# Chapter 8

# Community Warrant

I have suggested that today's professor undergoes a process of professional training and socialization that has few equals for its thoroughness, depth, and power. Even the most militantly individualistic and independent scholars[1] must normally go through an extensive professional formation within a discipline, including a probationary period as an assistant professor, followed by further demonstration of professional competence leading eventually to promotion to full professor. The scholar continues to practice his or her intellectual and scholarly activity within the confines of a larger disciplinary context that normally includes an academic department, disciplinary associations, journals, conferences, peer review for publications and grant funding, and so on.

Consider now how many and various institutional structures and institutionalized processes play crucial roles in our formation and continue to shape and constrain us as disciplinary specialists. Here's a very incomplete list: departments, graduate programs, graduate entrance examinations, letters of recommendation, undergraduate transcripts, graduate classes, general examinations ("Generals"), oral examinations ("Orals"), mentors and doctoral advisors, dissertations, dissertation defenses, national associations, "hiring fairs" at meetings of our discipline's national association, probationary faculty status, class visits and teaching evaluations by senior faculty, tenure and promotion committees, peer review for publishing and for grant funding and for selection of papers for national meetings. The list could easily be extended. These institutional entities—structures, statuses, processes, and the like—represent the authority of the discipline to *form* its members and assure that they *conform* to the internal goods and standards of the disciplinary community. The institutional expressions of a community of practice often embody those standards and rules and are equipped to maintain and enforce them.[2]

Many of these institutions act to separate the competent practitioner from the incompetent, and good scholarly work from shoddy. Take, for example, the practice of peer review, one of the hallmarks of academic professionalism (and professionalism generally). Faculty may be free to "do their own thing" in the library or the lab, but if they want to publish the results of their labors in a reputable professional journal, they will have to gain the blessing of their peers. This first step of "quality control" is often followed by others: reviews of the field in which a professional practitioner evaluates the current literature; citation indices that give some sense of influence (positive and negative) of a piece of scholarship on the larger field; evaluation of the faculty member's curriculum vitae as part of the process of competitive awarding of research grants; scrutinizing publications when deciding to hire, tenure, or promote. Again, the list could be easily extended.

## Philosophical Considerations

On the epistemological role of academic communities of practice—that is, on the ways in which academic communities warrant claims to knowledge—I want to make a complicated and potentially contentious point as simply (but some may feel, simplistically) as possible.

### "The Regulative Ideal of a Critical Community of Inquirers"

This is Richard Bernstein's label for a concept that goes back to the American Pragmatist Charles Sanders Peirce.[3] As we have seen, an academic community of practice is organized to establish and maintain its standards of excellence and to demand that its members aspire to achieving these standards in their pursuit of the goods that define the community. Several assumptions underlie these arrangements.

First, it is assumed that through a community's collective efforts, including peer review and other means of monitoring and evaluating individual work, a winnowing occurs that separates the better scholarship from the worse, the good from the bad, even perhaps the correct or "true" from the incorrect or "false." Some argue that under this winnowing process the community's understanding of its subject matter will converge over time to an approximation of the "truth of matters." This assumption rests on the conviction—challenged by some—that the convergence of many different, intimately intertwined arguments provides the most conclusive position.

Some even argue, more controversially still, that this convergence is the operational definition of the "truth of matters."[4] Scholars of this persuasion question to varying degrees whether there are objective standards for determining truth (and some also question whether truth is either a meaningful or useful concept apart from being a convenient shorthand for this community convergence).

Fortunately, for the purposes of this book, it is not necessary to take sides in this multifaceted dispute. It is sufficient to recognize that communities do *in practice* act as if the "truth of matters" is that which the community agrees to be the "truth of matters."

Second, disciplinary communities of practice also generally assume that their current results—constituting, perhaps, the "truth of matters" as the community currently understands it—may later prove to be incorrect as new evidence is found and new explanations or interpretations are developed. That is, the community operates on the assumption that its results are provisional, that at some later point the community may well converge on a different "truth of matters." The history of natural sciences in the last two centuries provides ample evidence of how our understanding of the truth of matters changes with the discovery of new evidence and the development of new theory.

Interestingly and importantly, the community may confidently assert the "truth of matters" as it understands it today and, by its lights, will be justified in making this assertion—that is, justified according to the community's internal standards and the evidence and argument offered to support its conclusion—yet prove to be wrong in the longer haul. More on this distinction between truth and justification in a bit.

Third, internal standards other than the "truth of matters"—however understood—are also frequently in play. In some cases, and with some disciplines, these other standards may play a more important role than any concern for the "truth of matters," however understood. Members of a discipline may be just as, or even more, concerned with other standards such as How plausible is the explanation or interpretation? (plausibility); How coherent is the overall picture it offers? (coherence, logical and otherwise); Does the explanation or interpretation suggest further applications or experiments? (fertility); How interesting or emotionally satisfying is the explanation or interpretation? (interest); Can it be generalized or even universalized? (generality and universalizability); How well does it fit with past results and past explanations or interpretations? (consistency); Is it simple, elegant, beautiful, ugly? (aesthetics); Is it in harmony with the overarching commitments of the community (fidelity).[5] And so on.

Many of these standards are subjective—they are in John Searle's terminology, "observer relative"[6]—yet not arbitrary. The goods, standards, and excellences internal to a disciplinary community may be difficult to grasp by someone outside the community, but they are generally public and, as such, can be evaluated and their application assessed. Members within the community have publicly available grounds for agreeing or disagreeing about how a particular interpretation or explanation is to be evaluated. With sufficient diligence an outsider to the community can also assess an interpretation or explanation in relation to the goods, standards, and excellences internal to the community. The public or intersubjective character of community standards does not, however, guarantee agreement. The choice and application of standards depends on practical reason, and individuals may legitimately disagree with each other on both choice and application. Judgments by both practitioners and outsiders have, in other words, an objective aspect, even though subjective inclination may also play a role. This is one of many reasons why communities are constituted in part by an ongoing debate on what its internal goods and standards should be.[7]

## Distinguishing between Truth and Justification

With this argument about how a community converges on the "truth of matters," I may have raised the specter of relativism, that is, some variant on the notion that Truth with a capital T is not an absolute and rather truth with a small t is relative to the perspective of a particular community. Certainly, some do understand the "regulative ideal of a critical community of inquirers" in this way. But this does not necessarily follow from the assumptions I have sketched.

If we distinguish between justification and truth, we may wish to assert that there exists a "Truth of matters" with a capital T. In so doing, however, we need to recognize that each attempt to *justify* such a truth claim relies for its justification on the goods and standards, the vocabulary and culture, the time and circumstances, of the community offering the justification. A particular community's truth may actually be Truth, but the community will not necessarily be able to convincingly justify its claim to those existing in a different community, with different goods and standards, different vocabulary and culture, different time and circumstances.

Truth may, in short, be absolute, but justification will be relative. Naturally, this claim itself may or may not be True, but my claim that it is true relies on a nonfoundational, fallibilistic, engaged pluralistic perspective that I find convincing from my and my community's "life situation" (*Sitz im Leben.*)[8] You will need to make up your own mind on the matter.

And again, for the purposes of establishing a basis for individual reflection or for conversation with faculty colleagues, it is not necessary for us to agree on this distinction between justification and the "truth of matters." It is sufficient to recognize that *in practice* when communities offer a justification for a claim, they may feel that given their goods and standards they are fully warranted in making their claim, yet at the same time members of a different community, with different goods and standards, may disagree and feel themselves fully justified in their disagreement. When we take a look at academic freedom in a later chapter, we'll see an example of how this works, and how in some circumstances community goods and standards intended to facilitate the search for truth may just as likely inhibit it.

## Religious or Spiritual Communities

As we discussed in an earlier session, religious or spiritual communities are also communities of practice with their own internal goods, standards, and practices. Much of the analysis developed in the previous section may apply to religious or spiritual communities as well. But even a moment's reflection suggests that religious or spiritual communities may differ from disciplinary communities regarding provisional claims, fallibility, justification, and standards of publicity.

Adherents to the various religious traditions of the world often claim that the truth of their faith is absolute and unchanging. Yet historians of that same tradition can point to many examples when beliefs and practices of the tradition appear to have changed or developed. Believers may explain this apparent discrepancy by tracing a line of "unchanging truth" through the vicissitudes of historical expressions. But to the outsider, and to many insiders as well (especially academics who are also insiders), the beliefs and practices of religious or spiritual communities appear in practice to have changed. What was at one time considered true has over time been superceded.[9] My point is a historical and sociological one: religious or spiritual claims appear also to be fallible and subject to change, although perhaps more slowly than is the case for disciplinary claims.

Normative differences—religious or spiritual communities claiming purchase on absolute, unchanging truth and disciplinary communities claiming that in principle everything in their discipline is subject to revision—may disguise greater similarities among the two. In my experience, religious or spiritual communities may be more open to revising even central claims than many adherents will admit, and disciplinary communities may be more resistant to revising central claims than many practitioners would like to believe. We need to be cautious when advancing any blanket generalizations about differences between disciplinary and religious or spiritual communities on the provisional nature of truth claims. The difference may be more in degree than in kind.

Another potential difference: some members of religious or spiritual communities seek to defend their beliefs by claiming that they belong, as theologian Ron Thiemann has remarked, to some "autonomous sphere wholly insulated from external scrutiny or critique."[10] Some academics take this claim at face value and then argue that this alleged independence from examination or critique disqualifies religious or spiritual perspectives from entering into academic discussion where all must be public and liable to scrutiny.

Yet if the community of practice model captures the phenomenon of religion at all well, the beliefs and practices of the community are certainly public. "To inquire concerning the faith of an individual or community," Thiemann explains,

> it is necessary to explore the set of practices within which the convictions of faith are displayed. To understand Christian notions of "love," for example, it would be helpful to read biblical texts (e.g. the parable of the Good Samaritan, the teachings on love in the Gospel and Epistles of John), to study theological treatises on the topic, and to learn about the benevolent practices of Christian communities across the centuries.[11]

Such a process, Thiemann continues, "is no more unusual or difficult than that which is required to understand a notion like 'freedom' in the American constitutional tradition" where one would have to scrutinize, say, the *Declaration of Independence* and *Constitution* and study writings such as *The Federalist* or Martin Luther King's *A Letter from a Birmingham Jail* and to look at the various practices overtime that embodied the American notion of freedom.[12] If publicity is the only issue, then most religious or spiritual beliefs and practices meet generally accepted standards. If religious or spiritual claims are to be properly ruled out of bounds in academic discussion, the decision should hang on the content of the belief and the warrants for accepting the belief, not on its publicity.[13]

## Conclusion

I have briefly sketched one way of thinking about how a disciplinary community judges scholarship and corrects errors and biases of individual members of the community. My account regarding community warrant obviously will not command agreement from all scholars. In an interesting piece of reflexivity, divergence on "the regulative ideal of a critical community of inquirers"[14] illustrates the very phenomenon I've been trying to get a handle on.

We scholars run the risk of misleading our students if we conceal how our individual scholarship depends on disciplinary communities to establish and maintain scholarly standards. If we fail to discuss the role of community in establishing what is taken to be true, persuasive, and consistent in scholarship, we also set our students (and perhaps ourselves) up to overlook the crucial communal element in the justification of any truth claims.

We also run the risk of overlooking how dependent on professional disciplinary communities is our notion of academic freedom—and how this dependence has some ironic implications. We turn to this issue in the next chapter.

# Chapter 9

# Academic Freedom

According to the understanding of academic freedom that is most common in American colleges and universities, it is a violation of a faculty member's academic freedom when individuals or institutions outside the faculty member's field attempt to constrain or dictate what individual faculty members research, publish, teach, or say extramurally. This is the sense of academic freedom most closely associated with the American Association of University Professors (AAUP). In matters of religion, it is a freedom that should shield faculty from the outside imposition of doctrinal beliefs and practices on research, publishing, teaching, and free speech as a citizen.

But is that all there is to academic freedom and religion? Ponder the following questions.

Is it a violation of a faculty member's academic freedom when colleagues within his discipline reject his article for publication because they find his arguments or use of evidence unconvincing or even repugnant, when they deny his application for a grant because they feel his area of research is insufficiently critical, or when they refuse to promote because of his allegiance to a particular school of thought? Is it a diminution of a colleague's freedom to choose her own research if foundations are willing to fund only certain types of research or if the job market prefers one subfield over another? Is it a violation of her academic freedom if colleagues criticize her for bringing explicitly religious perspectives into her teaching or scholarship even if she thinks the perspectives are germane?

Shift the focus to students and encounter more questions. If a faculty member should be accorded the freedom to teach, is there a correlative freedom for students to learn? Should a student be free from persistent intrusion of controversial materials that are not germane to the subject at hand? What if the material is germane but the student or outside observers see it as one-sided or prejudiced? Does it violate a student's academic freedom to learn when faculty espouse only one view of matters and either fail to mention or actively stigmatize alternate perspectives? What if the students' beliefs are held up to ridicule or peremptorily dismissed? Is it an abuse of a faculty member's academic freedom if he strongly advocates a view of the subject that conflicts with what the student has learned from family or religious community? Should the student be excused from learning material that he or she finds objectionable? Has the student's freedom to learn been compromised by the failure to include all "serious" alternative views?

And what of institutional academic freedom? Should colleges or universities be allowed to favor some approaches to research and teaching over others? Should church-related colleges or universities be allowed to discriminate in hiring or promoting on the basis of religious belief or practice? Is there an obligation for colleges and universities to provide "balance" and expose students to the "intellectual pluralism" found in many of the disciplines? What if such "balance" undermines beliefs deeply held by a sponsoring religious denomination?

Each question illustrates (more or less aptly) how complicated the matter of academic freedom may be when applied in practice. Whose academic freedom should have priority, and in what context? How does one balance between conflicting claims or determine to accede to one claim at the expense of another? Should institutions that decide one way be stigmatized as inferior to ones that decide another way?

In what follows, I shall concentrate on how these distinctions may bear on religion on our campuses. But it should be immediately apparent that in today's debates religious questions very much take the back seat to political ones. In thinking about how you might balance or resolve conflicting claims regarding academic freedom and religious convictions, you should consciously substitute "political" for "religious" and see whether your position changes, and ask yourself why or why not.

## Academic Freedom and Disciplinary Communities of Practice

As historian Thomas Haskell reminds us, academic freedom as we know it in today's colleges and universities arose from the same process that gave rise to today's disciplinary communities of practice.[1] In the chapter "Disciplinary Formation," we consider how academic disciplines arose gradually in the late nineteenth and early twentieth centuries out of a process of specialization and differentiation. These developments were fueled by a confidence in science and embodied the conviction that advances in understanding and knowledge were necessarily the progressive achievements of communities of professional scholars.

Self-regulation is a professional hallmark. As we saw in "Community Warrant," self-regulation was also thought to move a disciplinary community toward a better understanding of their subject matter. Through its collective efforts, including peer review and other means of monitoring and evaluating individual work, a community developed its best understanding of its subject matter. Of course, today's best understanding might have to yield to tomorrow's better evidence or new explanation. But it took a community to accomplish this crucial winnowing.

On this understanding of the collective nature of good scholarship, our disciplinary forebears made bold to claim that authority to pronounce on matters within a field belonged exclusively to those of demonstrated competence who were answerable to the disciplinary community. Competence was acquired and demonstrated by undergoing university training and certification, securing professional appointment, and regularly producing research accepted by disciplinary peers.

The American conception of academic freedom arises out of this process of disciplinary formation with its understanding of earned intellectual authority. It relies on the notion of disciplinary communities of practice and specialized expertise in service to society to explain and justify the privileges and responsibilities it conveys. This appears clearly from even a brief perusal of the 1915 *General Declaration of Principles* issued at the founding of the American Association of University Professors, the umbrella community of practice that embraced all the constituted disciplinary communities to which university and college teachers belonged. The *Declaration* was written by economist Edwin R. A. Seligman and was substantially revised by historian Arthur Lovejoy.[2]

Under the heading "The Nature of the Academic Calling," Seligman and Lovejoy lay out "the chief reasons, lying in the nature of the university teaching profession, why it is to the public interest that the professional office should be one both of dignity and of independence." Under the joint assumptions that education is "the cornerstone of the structure of society" and that "progress in scientific knowledge is essential to civilization," Seligman and Lovejoy argue for the importance of attracting "men of the highest ability, of sound learning, and of strong and independent character" into a profession that will not reap large pecuniary rewards. To this end, they must be assured "an honorable and secure position" and granted the "freedom to perform honestly and according to their own consciences the distinctive and important function which the nature of the profession lays upon them." Independence is the reward for service to society.[3]

The service or function of the scholarly professional "is to deal at first hand, after prolonged and specialized technical training, with the sources of knowledge; and to impart the results of their own and of their fellow-specialists' investigation and reflection, both to students and to the general public, without fear or favor." To discharge this function properly, the university teacher must "be exempt from any pecuniary motive or inducement to hold, or to express, any conclusion which is not the genuine and uncolored product of his own study or that of fellow-specialists." In fact, for the professoriate to do its proper work, universities' needs must be "so free that no fair-minded person shall find any excuse for even a suspicion that the utterances of university teachers are shaped or restricted by the judgment, not of professional scholars, but of inexpert and possibly not wholly disinterested persons outside of their ranks."[4]

Why must teachers be beyond suspicion? In explaining this necessity, Seligman and Lovejoy suggest that it is actually the lay public who employs the "scientific experts" who are "trained for, and dedicated to, the quest for truth," and not those who manage or endow the universities that pay their salaries. "The lay public is under no compulsion to accept or to act upon the opinions of the scientific experts whom, though the universities, it employs," they write, "But it is highly needful, in the interest of society at large, that what purport to be the conclusions of men trained for, and dedicated to, the quest for truth, shall in fact be the conclusions of such men, and not echoes of the opinions of the lay public, or of the individuals who endow or manage universities."[5]

To degree that faculty are, or are thought to be, "subject to any motive other than their own scientific conscience and a desire for the respect of their fellow-experts"—here we see the regulative ideal of the community of practice—their profession, Seligman and Lovejoy conclude, is corrupted with the result that "its proper influence upon public opinion is diminished and vitiated; and society at large fails to get from its scholars, in an unadulterated form, the peculiar and necessary service which it is the office of the professional scholar to furnish."[6]

University trustees appoint faculty, but faculty are not their employees. "For, once appointed, the scholar has professional functions to perform in which appointing authorities have neither competency nor moral right to intervene." Seligman and Lovejoy explain that the university teacher is responsible "primarily to the public itself, and to the judgment of his own profession." While the teacher is responsible "with respect to certain external conditions of his vocation" to the university that employs him, "in the essentials of his professional activity his duty is to the wider public to which the institution itself is morally amenable." Seligman and Lovejoy offer an analogy with the appointment of federal judges. "University teachers," they explain, "should be understood to be, with respect to the conclusions reached and expressed by them, no more subject to the control of the trustees than are judges subjects to the control of the President with respect to their decisions." Nor should trustees be held responsibility for what professors say, just as the president is not expected always to agree with the judgments reached by those he appoints. Trustees play an essential and honorable role in a university, but faculties "hold an independent place, with quite equal responsibilities—and in relation to purely scientific and educational questions, the primary responsibility."[7] It is worth noting that with this argument, Seligman and Lovejoy are deploying only an lightly secularized notion of vocation that we explored in "Narrative Identity."

When academic freedom is seen, as Seligman and Lovejoy saw it, as the condition that enables the professional disciplinary communities of practice to serve the larger society, it helps us understand why it would be deemed a violation of the professional freedom and concomitant responsibilities that belong exclusively to the disciplinary community of practice to propose that any party or institution outside the disciplinary communities of practice (or the colleges or universities where disciplinary specialists teach and do research) should have authority to set, judge compliance, or enforce any of these listed professional obligations. In 1915, these claims were controversial and contested. Twenty-five years later, they had swept the field.

## Faculty Academic Freedom

The 1940 *Statement of Principles on Academic Freedom and Tenure* is the paradigmatic expression of the notion of academic freedom within the United States.[8] It was originally agreed upon by the representatives of the American Association of University Professors (AAUP) and of the Association of American Colleges (since 1995, the Association of American Colleges and Universities) and subsequently endorsed by more than 150 academic associations or societies and recognized (with limitations) by the American courts.[9]

The 1940 *Statement* recognizes four kinds of freedom that together comprise the notion of faculty academic freedom in America: the freedom to teach, research, publish, and speak as individual citizens. To begin, the *Statement* claims that teachers are entitled to "full freedom" in research and in the publication of their results. The only limitation, if any, arises when the research is for "pecuniary return," in which cases an "understanding with the authorities of the institution" should be obtained.

Teachers are also entitled to "freedom in the classroom in discussing their subject" but the *Statement* goes on to say that teachers "should be careful not to introduce into their teaching controversial matter which has no relation to their subject." In a subsequent 1970 *Interpretive Comments*, the AAUP explains that the intent was not to discourage discussions of the controversial in classroom setting. "Controversy," they explain, "is at the heart of the free academic inquiry which the entire statement is designed to foster." Rather the passage "serves to underscore the need for teachers to avoid persistently intruding material which has no relation to their subject."

In the article dealing with "freedom in the classroom," the 1940 *Statement* recognizes that there may be some limitations of academic freedom "because of religious or other aims of the institution," but it specifies that such limitations should be clearly stated in writing at the time of appointment. Although this limitation is included in the article dealing with classroom teaching, the AAUP views this provision as applying to all the four elements of academic freedom.[10] In the subsequent *Interpretive Comments* of 1970 the AAUP states that "[m]ost church-related institutions no longer need or desire the departure from the principle of academic freedom implied in the 1940 *Statement*, and we do not now endorse such a departure." This *Interpretive Comment* raises some serious questions for church-related colleges and universities, which we'll address in the section "Institutional Academic Freedom."

Finally, the *Statement* recognizes that faculty are also citizens, members of a "learned profession," and officers of an educational institution. With these overlapping roles in mind, the *Statement* indicates that when faculty "speak or write as citizens, they should be free from institutional censorship or discipline, but their special position in the community imposes special obligations." Even when speaking as citizens, faculty should be mindful that the public "may judge their profession and their institution by their utterances." For this reason, faculty "should at all times be accurate, should exercise appropriate restraint, should show respect for the opinions of others, and should make every effort to indicate that they are not speaking for the institution." Subsequent commentary strongly suggests that the AAUP believes that the exercise of the right to free speech enjoyed by a faculty member as a citizen should be grounds for dismissal only if it clearly demonstrates his or her unfitness for the position. Even then, the faculty member's entire record as a scholar and teacher should be taken into account.

## Freedom From and Freedom For

In "Two Concepts of Liberty (1958)," philosopher Isaiah Berlin explores the difference between what he terms "negative" and "positive" concepts of freedom.[11] Negative freedom is freedom *from* outside interference in the pursuit of one's goals. Positive freedom is the freedom *for* self-realization, a freedom that allows "my life and decisions to depend on myself, not on external forces of whatever kind." I shall be adapting these distinctions to make a point about today's American version of faculty academic freedom.

## Negative Freedom and Professional Self-Regulation

Negative freedom when applied to the academic realm entails the right to pursue one's scholarship and teaching free from outside interference, that is, free from the interference of administrators, trustees, politicians, or other guardians of public orthodoxy.[12] The key qualifier here is "outside." Recall that disciplinary communities of practice are *professions*, and one of the hallmarks of professional communities is *self-regulation*. In this regard, disciplinary communities are paradigmatic professional communities. The implicit social contract that underlies academic freedom runs something like this: "In consideration for services rendered to the larger society, and in deference to the expertise within the disciplinary community, the larger society will not interfere in scholarship or teaching so long as the disciplinary community itself provides adequate self-regulation." We'll look at the social quid pro quo in this implicit contract in a moment; note for now, however, that freedom from outside interference involves considerable constraints on freedom from within the disciplinary guild.

We can see this outside–inside distinction more clearly if we consider the constraints under which probationary faculty members (i.e., instructors or assistant professors) operate. They may be free from demands made from outside the faculty—demands, for example, to hew to some orthodoxy whether scholarly, religious, political, or economic—but they would be ill-advised professionally to ignore the expectations and preferred orthodoxies of those senior disciplinary colleagues on whom their future at the institution and within the profession rests. While they are likely to have some leeway regarding internal orthodoxies, and especially so if they possess a large degree of creative intelligence and more than a little bravado, they nonetheless remain subject to the judgment of their senior peers, a judgment expressed in everything from doctoral exams to peer reviews and evaluations for promotion and tenure.

In principle, academic freedom means not only freedom from interference by administrators or trustees but also freedom from interference by colleagues outside one's own discipline and department. Each discipline has its own expertise, and only one's expert peers should stand in judgment of junior professionals. In practice this freedom from interference from colleagues outside one's discipline is at best a qualified right. Most colleges and universities employ some form of college- or university-wide promotion and tenure committee where members of many different departments review the recommendations of specific departments. Not uncommonly, a unanimous recommendation for promotion and tenure from, say, the Sociology Department may be overruled by nonsociologists on the university- or college-wide Tenure and Promotion Committee.

In practice, then, there are degrees of academic freedom: academic freedom is greatest vis-à-vis authorities outside the faculty and disciplinary community (e.g., vis-à-vis administrators, trustees, alums, public officials, or religious authorities), moderate vis-à-vis faculty in other disciplines and departments within the college or university, and most circumscribed vis-à-vis peers in one's own discipline and department. To put this in terms of competency and hence legitimacy, the principles of academic freedom dictate that those outside academe, who lack the professional competency, have little or no legitimate claim to judge the products of scholarship and teaching; university or college colleagues outside one's department may legitimately review the application of standards set by a particular department and discipline; and members of the discipline and department may legitimately set the standards and evaluate the degree to which a colleague has met those standards. This system is not immune from abuse, but on balance, it works reasonably well.

Regrettably, but understandably, orthodoxies within the discipline can in some cases be as harmful to truly independent and meritorious scholarship and teaching as orthodoxies from without. As we discussed in "Cautionary Tales," for example, internal orthodoxies regarding the "type" of person fit to teach the ideals and values of Western civilization (they assumed this was equivalent to Christian civilization of a certain sort) kept Jews and Catholics out of the ranks of the humanities departments well into the mid-twentieth century. More recently, internal orthodoxy regarding the appropriateness of even acknowledging religious or analogous motives in scholarship and teaching may have tended to disadvantage scholars who were religious and encouraged them to keep their religious views private. One enjoys academic freedom vis-à-vis outside interests only if one is a member of the disciplinary profession, and the price of that membership is reasonable fidelity to the goods and standards of the profession itself. If these goods and standards argue against even discussing the bearing of religious motives on research or teaching, appeals to academic freedom will be of little avail.

No one today would wish to countenance the discrimination that Jews, Catholics, and other religious minorities faced during the first five or six decades of the twentieth century simply because of their religious status. But the matter may be more difficult to sort out when professions discriminate against particular religious beliefs that run counter to professional consensus. To take an extreme but pertinent example, academic freedom was explicitly meant to allow biology professors at church-related colleges to teach evolution despite objections from denominational authorities. But academic freedom may not allow a biology teacher to teach "creation science." This may look like a double standard, but it simply reflects the inside–outside distinction that is central to academic freedom and professional disciplines. The social contract underlying academic freedom gives to the profession not only the right but also the responsibility to establish and uphold the goods and standards internal to the disciplinary practice. The biological discipline considers evolutionary theory as good science, even as it continually argues about and refines the details of that theory. In contrast, it considers "creation science" as bad science, or rather as no science at all. Sincere religious conviction has little or no purchase on this professional judgment.

Even this distinction requires some nuance. Orthodoxies within the disciplines come with varying degrees of warrant. In biology, evidence for evolution and evolutionary processes is unusually broad and deep. By way of contrast, evidence for some theories of evolutionary psychology is relatively scant and the associated theories are highly contested within the broader discipline. Where the warrant is considerable, freedom may be constrained; where warrant is slight and evidence conflicting, there is greater freedom to disagree. One may argue about the specifics in each case—I happen to agree with the biology profession on the matter of "creation science"; you may not—but it is important to realize that academic freedom from outside forces is purchased with restraints within the profession.

## Positive Freedom and Religion

The concept of academic freedom partakes largely of what Berlin termed "negative" freedom, that is, freedom *from* outside interference. But encroachments on scholarship and teaching from the side of various religious traditions often adduce variants on Berlin's "positive" freedom, that is, the freedom *for* what is deemed a person's "true" self-determination. The positive sense of the word "liberty," Berlin explains,

> derives from the wish on the part of the individual to be his own master. I wish my life and decisions to depend on myself, not on external forces of whatever kind. I wish to be the instrument of my own, not of other men's, acts of will. I wish to be a subject, not an object; to be moved by reasons, by conscious purposes, which are my own, not by causes which affect me, as it were, from

> outside. I wish to be somebody, not nobody; a doer—deciding, not being decided for, self-directed and not acted upon by external nature or by other men as if I were a thing, or an animal, or a slave incapable of playing a human role, that is, of conceiving goals and policies of my own and realising them.[13]

Such aspirations may seem unexceptional, but the danger lies, as Berlin makes eloquently clear, in what one understands by "self-mastery," and especially when self-mastery means finding one's "true" or "higher" self. It is at this point that a religious understanding of self-mastery may seem to some to be oppressive rather than liberating. And when the "self" takes on a collective cast, the opportunities for tyranny abound. Here's Berlin again:

> Have not men had the experience of liberating themselves from spiritual slavery, or slavery to nature, and do they not in the course of it become aware, on the one hand, of a self which dominates, and, on the other, of something in them which is brought to heel? This dominant self is then variously identified with reason, with my "higher nature," with the self which calculates and aims at what will satisfy it in the long run, with my "real," or "ideal," or "autonomous" self, or with my self "at its best"; which is then contrasted with irrational impulse, uncontrolled desires, my "lower" nature, the pursuit of immediate pleasures, my "empirical" or "heteronomous" self, swept by every gust of desire and passion, needing to be rigidly disciplined if it is ever to rise to the full height of its "real" nature. Presently the two selves may be represented as divided by an even larger gap; the real self may be conceived as something wider than the individual (as the term is normally understood), as a social "whole" of which the individual is an element or aspect: a tribe, a race, a Church, a State, the great society of the living and the dead and the yet unborn. This entity is then identified as being the "true" self which, by imposing its collective, or "organic," single will upon its recalcitrant "members," achieves its own, and therefore their, "higher" freedom.[14]

Such a concern for the "true" self and the "higher" freedom may undergird some attempts by religious bodies to regulate what is taught within denominational colleges and universities. To use Berlin's distinction, the religious community may believe that the (positive) freedom of students to realize their "true, God-given human nature" should trump scholars' (negative) freedom to teach whatever they and their disciplinary community of practice think accords with the disciplinary community of practice's internal goods and standards.

I mention this not to defend this positive perspective but only to explain a possible motivation of some of academic freedom's opponents. In today's diverse world, Berlin's concluding sentiment strikes me as the most prudent for twenty-first-century higher education. "Pluralism, with the measure of 'negative' liberty that it entails," Berlin wrote,

> seems to me a truer and more humane ideal than the goals of those who seek in the great disciplined, authoritarian structures the idea of "positive" self-mastery by classes, or peoples, or the whole of mankind. It is truer, because it does, at least, recognise the fact that human goals are many, not all of them commensurable, and in perpetual rivalry with one another.[15]

And, of course, that pluralism may also be expressed in a diversity of institutions of higher education, including at least a few that advocate the (positive) freedom of students to realize their "true, God-given human nature" and accordingly insist on that institutional academic freedom trump faculty academic freedom at least in some domains. We now turn to student and institutional academic freedom, which complicates matters further.

## Student Academic Freedom

In the same famous piece on "Two Concepts of Liberty (1958)" we drew from earlier, Isaiah Berlin advances a thesis centrally associated with his work as a political philosopher, namely, that in the world of ordinary experience "we are faced with choices between ends equally ultimate, and claims equally absolute, the realisation of some of which must inevitably involve the sacrifice of others."[16] As he put it elsewhere with customary eloquence, "Some among the Great Goods cannot live together. That is a conceptual truth. We are doomed to choose, and every choice may entail an irreparable loss."[17] Take liberty and equality. "Both liberty and equality are among the primary goals pursued by human beings through many centuries," Berlin explains,

> but total liberty for wolves is death to the lambs, total liberty of the powerful, the gifted, is not compatible with the rights to a decent existence of the weak and the less gifted. . . . Equality may demand the restraint of the liberty of those who wish to dominate; liberty—without some modicum of which there is no choice and therefore no possibility of remaining human as we understand the word—may have to be curtailed in order to make room for social welfare, to feed the hungry, to clothe the naked, to shelter the homeless, to leave room for the liberty of others, to allow justice or fairness to be exercised.[18]

It is debatable whether academic freedom should be listed among the "Great Goods," but Berlin's insight about the clash of goods and the necessity of choice would seem to apply if one assumes that students enjoy an academic freedom to learn that is equivalent to the faculty's freedom to teach. The academic freedom of faculty to choose what they teach (and do not teach) could be seen as infringing the academic freedom of students to learn all that they wish to (or should) learn. Total freedom for faculty could well mean intellectual servility for students.

But within the American context this is not the case, and for good reason related to history we just covered. The American concept of academic freedom arises out of the historic development of disciplinary communities of practice. It relies on a particular understanding of how the (approximate, fallible, provisional) "truth of matters" is arrived at by the ongoing work of a specialist, professional community. It is a professional privilege offered in exchange for service to the larger society.

Students are at best junior apprentices within disciplinary communities of practice, and most do not even gain this status until graduate school. They are the recipients of the findings of the disciplinary communities, not coequal partners in their discovery. They are not professionals. And their period of education is meant to equip them for whatever service they will do later for a society that values the knowledge and skills and judgment that they will acquire.

When we see student academic freedom within this context, it helps us to see why "student academic freedom" appears so limited in relation to faculty academic freedom. Student academic freedom is a derived freedom, arising out of the professional obligation of members of an academic discipline. It is not a freedom independent of faculty academic freedom, much less a freedom coequal with faculty academic freedom. Keeping this in mind will help make sense of what follows.

## 1915 Declaration of Principles

From its founding in 1915, the AAUP has focused its efforts and attention on defining, justifying, and securing faculty freedom of inquiry, publication, and teaching, and freedom of extramural speech and action. Since then it has also recognized that the freedom of the faculty member to teach entails certain correlative obligations to secure students' freedom to learn. In the AAUP's 1915 *Declaration of Principles*, some of these correlative obligations are spelled out in more detail than is the case in the 1940 *Statement* or 1970 *Interpretive Comments* and with more emphasis on what the faculty member should *do* as opposed to *not do*.[19] That is, it spells out in greater detail what its authors think teachers should do as well as refrain from doing.

The 1915 *Declaration* states that the "university teacher, in giving instructions upon controversial matters, while he is under no obligation to hide his own opinion under a mountain of equivocal verbiage, should, if he is fit in dealing with such subjects, set forth justly, without suppression or innuendo, the divergent opinions of other investigators." In other words, teachers need not hide their own opinions but if they are competent to do so, they should fully and fairly present the differing scholarly opinions, or at least, as the *Declaration* puts it, "the best published expressions of the great historic types of doctrine upon the questions at issue." Above all the teacher should remember "that his business is not to provide his students with ready-made conclusions, but to train them to think for themselves, and to provide them access to those materials which they need if they are to think intelligently."[20]

But who decides which "divergent opinions of other investigators" should be set forth or what constitutes "the best published expressions" on controversial issues? The *Declaration* is clear on this matter. It is "inadmissible," the *Declaration* states, "that the power of determining when departures from the requirements of the scientific spirit and method have occurred, should be vested in bodies not composed of members of the academic profession." Why? Because such external bodies lack the competence and their intervention will always lie under the suspicion of being "dictated by other motives than zeal for the integrity of science." The responsibility, therefore, lies with university teachers themselves. It may be difficult for the profession, the *Declaration* admits, but this responsibility cannot be evaded. And in words remarkably prescient given current controversies we examine in the next section, the *Declaration* goes on to say that if the academic profession

> should prove itself unwilling to purge its ranks of the incompetent and the unworthy, or to prevent the freedom which it claims in the name of science from being used as a shelter for inefficiency, for superficiality, or for uncritical and intemperate partisanship, it is certain that the task will be performed by others—by others who lack certain essential qualifications for performing it, and whose action is sure to breed suspicions and recurrent controversies deeply injurious to the internal order and the public standing of universities.[21]

The AAUP's 1915 founding *Declaration* also recognizes an obligation of teachers to exercise "certain special restraints" when instructing "immature students" especially in their first two years, when, as the *Declaration* delicately puts it, "the student's character is not yet fully formed, [and] his mind is still relatively immature." The concern here seems to be that the instructor present "scientific truths" with discretion and gradually "with some consideration for the student's preconceptions and traditions, and with due regard to character-building." The *Declaration* continues,

> The teacher ought also to be especially on his guard against taking unfair advantage of the students' immaturity by indoctrinating him with the teacher's own opinions before the student has had an opportunity fairly to examine other opinions upon the matters of question, and before he has sufficient knowledge and ripeness in judgment to be entitled to form any definitive opinion of his own. It is not the least service which a college or university may render to those under its instruction, to habituate them to looking not only patiently but methodically on both sides, before adopting any conclusion upon controverted issues.[22]

The point of it all, it seems, is to wean students gradually and with due consideration for their backgrounds and immaturity from their "preconceptions and traditions" (if that is what "scientific truth" requires) and to get students to reach the point where they can reason and think about such issues on their own. With this advice, the authors of the *Declaration* did "not intend to imply that it is not the duty of an academic instructor to give to any students old enough to be in college a genuine intellectual awakening and to arouse in them a keen desire to reach personally verified conclusion upon all questions of general concernment to mankind, or of special significance for their own time." "It is better for students to think about heresies than not to think at all," wrote the president of Reed College, who is quoted with approval in the *Declaration.* The advice in this section goes simply to *how* such "intellectual awakening" should be brought about, namely, with "patience, considerateness, and pedagogical wisdom."[23]

## The 1940 *Statement of Principles on Academic Freedom and Tenure* and Forward

Twenty-five years later, the AAUP's primary focus remains squarely on defining, justifying, and securing academic freedom for faculty. But the association also continues to recognize that with this freedom comes responsibilities to secure the students' correlative freedom to learn.

First, as noted earlier, the 1940 *Statement* together with the 1970 *Interpretive Comments* states that faculty should not introduce into their teaching "controversial matter which has no relation to their subject." The stress is placed on the lack of pertinence. Controversy itself is, as the *Interpretive Comment* puts it, "at the heart" of free academic inquiry. At issue is the need for faculty "to avoid persistently intruding material which has no relation to their subject."[24] If faculty fail in this obligation, one may infer that they may be misusing their academic freedom to teach at the expense of their students' academic freedom to learn.

Second, in 1967, representatives of the AAUP and several student, student-service, and education associations[25] issued a *Joint Statement on Rights and Freedoms of Students*. The primary stress, as might be expected from the 1940 *Statement*, is on securing conditions conducive to the freedom to learn, for students and for all other members of the academy. Two points in this *Joint Statement* are particularly pertinent to students' right to learn. Point one, faculty are enjoined to encourage free discussion, inquiry, and expression in the classroom and in conference. "Students," the *Joint Statement* submits, "should be free to take reasoned exception to the data or views offered in any course of study and to reserve judgment about matters of opinion," adding a significant qualifier that stresses that the student's freedom is a freedom *to learn*, that students are nonetheless "responsible for learning the content of any course of study for which they are enrolled." Point two, the *Joint Statement* also states that student performance "should be evaluated solely on an academic basis, not on opinions or conduct in matters unrelated to academic standards" and that students "should have protection through orderly procedures against prejudiced or capricious academic evaluation." Again, this norm is qualified in a way that points to the student's freedom *to learn*: "At the same time, they [students] are responsible for maintaining standards of academic performance established for each course in which they are enrolled."[26] Naturally, such procedural safeguards should afford students a chance to appeal evaluations that the students feel were made on the basis of religious or political belief as well as characteristics such as race, ethnicity, gender, or sexual orientation.

The 1940 *Statement* and the 1967 *Joint Statement* speak sparingly of faculty or student *rights*, and with regard to instruction, only of the right of faculty to freedom to teach and the right of students to freedom to learn. Faculty also enjoy the right of free speech, as do students. Even in the *Joint Statement*, with its title referring to the "rights and freedoms of students," student rights are evoked only in the section on "off-campus freedom of students" dealing with student rights as citizens and rights students should enjoy procedurally in disciplinary proceedings. In the footnotes, students are also said to have a "right" to be informed about "the institution, its policies, practices, and characteristics," and a "right to be free from discrimination on the basis of individual attributes not demonstrably related to academic success in the institution's programs, including, but not limited to, race, color, gender, age, disability, national origin, and sexual orientation." The emphasis, then, rests largely on the right to freedom, for faculty to teach and for students to learn.

To sum up, the academic freedom of faculty includes correlative responsibilities to students which, from the AAUP's perspective, secure for students freedom to learn. These flow naturally from an understanding of higher education that sees it comprising members of professional disciplinary communities of practice who have local responsibilities but are primarily answerable to their peers in performing professional service to the larger society. Teaching students is part of that service and entails a variety of professional obligations. Student academic freedom consists largely in what a conscientious exercise of the obligations would supply.

First, controversial matters that are not pertinent to the subject under study freedom should not be intruded into classroom or conference meetings. This professional responsibility does not require the teacher to avoid all controversial issues. On the contrary, controversy is, after all, one of the drivers by which the disciplinary community separates over time the better scholarship from the worse, the closer approximation to the truth of matters from the more distant. But in those cases where a member of the community fails to heed this professional responsibility, or at least is accused of failing to heed it, the professional community and the local colleges and universities need to develop policies and provide mechanisms for hearing complaints and restoring proper professional practice.

Second, academic performance should be evaluated strictly on the basis of academic standards and practices of the discipline. When evaluating student performance, it would be unprofessional for teachers to employ nonacademic and discriminatory considerations such as politics, religion, race, gender, or class. The disciplinary community (and the college or university where the disciplinary specialist teaches) has the responsibility to assure that such professional standards are followed, and so it must offer policies and procedures by which a student may appeal an evaluation of academic performance that the student feels has been tainted by nonprofessional considerations.

Third, disciplinary professionals owe their students an unreserved presentation of their own perspective on issues in dispute within their field. But good professional practice should also include a fair and balanced presentation of divergent opinions. As a certified member of the discipline, the teacher has the authority to decide what the divergent opinions are that are worthy of presentation. The presumption is that a professional will be (as far as humanly possible) fair in his choice and will not capriciously or dishonestly omit or distort those opinions of other recognized disciplinary scholars with whom he or she disagrees. To address cases of unprofessional conduct in this regard, the disciplinary community of practice, and the colleges and universities in which individual disciplinary scholars teach and research, has the professional obligation to issue and maintain policies that ensure professional behavior in this regard. But consistent with its understanding of the intellectual authority held solely by the disciplinary community of practice, all judgments of what properly constitutes the diverge opinions lie ultimately with the community itself.

## The Call for "Intellectual Diversity"

As we have seen, the authors of the 1915 *Declaration* were at pains to suggest that teachers need not hide their own point of view but should nonetheless fully and fairly present the "divergent opinions of other investigators." The business of a teacher "is not to provide his students with ready-made conclusions, but to train them to think for themselves, and to provide them access to those materials which they need if they are to think intelligently." But it qualified this obligation with a proviso we'd expect given the special role Seligman, Lovejoy, and their fellows in the founding of the AAUP attributed to disciplinary communities of practice, namely, that it is "inadmissible that the power of determining when departures from the requirements of the scientific spirit and method have occurred, should be vested in bodies not composed of members of the academic profession."[27] In the early twenty-first century, this assertion has come under direct challenge by self-proclaimed champions of student academic freedom and especially of what they term "the principle of intellectual diversity."

This challenge is being led by the conservative activist David Horowitz and his Students for Academic Freedom (SAF) to "end the political abuse of the university and to restore integrity to the academic mission as a disinterested pursuit of knowledge."[28] To this announced end, SAF and its allies have negotiated with colleges and universities, both public and private, and have urged the adoption of legislation at both the national and the state level. At the heart of this campaign is what is termed "the principle of intellectual diversity." The promotion of "intellectual diversity" on campus is the first goal in the SAF mission statement,[29] and this goal is mentioned in the declarations and findings of most of the legislation that has been introduced nationally and in several states.[30] It is repeated regularly in the model Joint Resolution offered by SAF.[31] It is regularly paired with academic freedom.[32] It lies at the heart of SAF's proposed *Academic Bill of Rights* and the associated *Student Bill of Rights*.

What does "intellectual diversity" mean to the SAF? In explaining its goals, the SAF's Mission Statement states that "the atmosphere that prevails on most college campuses today does not foster intellectual diversity or the disinterested pursuit of knowledge. Liberal Arts faculties at most universities are politically and philosophically one-sided, while partisan propagandizing often intrudes into classroom discourse." This bias, the Mission Statement explains, "is reflected in the curriculum of courses available, in the manner in which they are taught, in readings assigned for classroom study, and in discussions only open to one side of a debate." The SAF seeks to overcome that bias. Its goal is "to secure greater representation for under-represented ideas and to promote intellectual fairness and inclusion in all aspects of the curriculum, including the faculty hiring process, the spectrum of courses available, reading materials assigned, and in the decorum of the classroom and the campus public square."[33]

To this end—"to secure the intellectual independence of faculty and students and to protect the principle of intellectual diversity"—the *Academic Bill of Rights* offers eight principles or procedures. I shall single out three to illustrate the thrust behind SAF's principle of "intellectual diversity."

4. Curricula and reading lists in the humanities and social sciences should reflect the uncertainty and unsettled character of all human knowledge in these areas by providing students with dissenting sources and viewpoints where appropriate. While teachers are and should be free to pursue their own findings and perspectives in presenting their views, they should consider and make their students aware of other viewpoints. Academic disciplines should welcome a diversity of approaches to unsettled questions.
5. Exposing students to the spectrum of significant scholarly viewpoints on the subjects examined in their courses is a major responsibility of faculty. Faculty will not use their courses for the purpose of political, ideological, religious or anti-religious indoctrination.
8. Knowledge advances when individual scholars are left free to reach their own conclusions about which methods, facts, and theories have been validated by research. Academic institutions and professional societies formed to advance knowledge within an area of research, maintain the integrity of the research process, and organize the professional lives of related researchers serve as indispensable venues within which scholars circulate research findings and debate their interpretation. To perform these functions adequately, academic institutions and professional societies should maintain a posture of organizational neutrality with respect to the substantive disagreements that divide researchers on questions within, or outside, their fields of inquiry.[34]

At first blush, these sentiments will strike most readers as unexceptional or even admirable. Yet each carries within it what Seligman and Lovejoy would have termed the "inadmissible" remedy for professional conduct that "the power of determining when departures from the requirements of the scientific spirit and method have occurred, should be vested in bodies not composed of members of the academic profession."[35]

Take point 4: Curricula and reading lists in the humanities and social sciences are expected to provide students with "dissenting sources and viewpoints where appropriate." Who determines what "dissenting sources and viewpoints" should be included in the curriculum or on reading lists? The faculty member or some outside authority? Or point 5: Who determines whether faculty have met their "major responsibility" "to expose students to the spectrum of significant scholarly viewpoints on the subjects in their course"? Who decides how wide the "spectrum of significant scholarly viewpoints" should be? Or point 8: Under a professional understanding of academic freedom, it would be the scholarly societies and their practice of peer review that decides whether a question is "unsettled" (point 4) and what constitutes "substantive disagreements that divide researchers on questions within, or outside, their field of inquiry." But the *Academic Bill of Rights* would have professional societies and associations "maintain a posture of organizational neutrality" regarding such matters. Who then, if not professional peers, should decide?

Here is how one legislator thought to settle the matter. In Florida, representative Robert Baxley introduced the Florida House Bill 837 that followed closely the model legislation advocated by SAF. Florida Senate staff analysis suggested that the bill, if passed (it died in committee on May 6, 2005), appeared to "create a cause of action for students to litigate against the public postsecondary education institution in which they are enrolled" if the student felt that his or her "rights" as spelled out in the bill had been violated.[36] These rights included the "right to expect a learning environment in which they will have access to a broad range of serious scholarly opinion pertaining to the subjects they study," and the bill continued under this article, "In the humanities, the social sciences, and the arts, the fostering of a plurality of serious scholarly methodologies and perspectives should be a significant institutional purpose." When asked about this possibility, representative Baxley reportedly responded, "Being a businessman, I found out you can be sued for anything. Besides, if students are being persecuted and ridiculed for their beliefs, I think they should be given standing to sue."[37] In a radio exchange with a faculty opponent to the legislation, he clarified a remark attributed to him that he thought a student could sue if faculty refused to even listen to arguments about Intelligent Design:

> First of all, the whole idea of intelligent design being taught is never something that I have advocated. I merely illustrated that I went on an anthropology class as a student and was dogmatically told that evolution is a fact. There's no missing link. I don't even want to hear anything about creation or intelligent design. And if you don't like any of that, there's the door. That kind of dogmatism is what I was addressing, not that they needed

> to teach—they can teach whatever they want to teach, but what the bill requires is that you give different schools of thought and not just the dogma of an individual professor.[38]

It would seem at least in representative Baxley's understanding of his proposed legislation that a "dogmatic" insistence that Intelligent Design is not a science could invite a student law suit.[39]

SAF's version of student academic freedom clashes with the AAUP's version of faculty academic freedom. More of one means less of the other. The traditional American version of academic freedom for professional members of disciplinary communities may be undermined by "the principle of intellectual diversity,"[40] for the principle, if secured by legislation, transforms a professional responsibility on the part of faculty into a student right that may even be enforceable in court.[41]

## Student Academic Freedom and Religion

If wc assume that the traditional American concept of academic freedom as primarily a faculty right (with correlative obligations to students) will prevail at least for now, then the observations offered earlier about the difference between external and internal orthodoxies applies doubly to students. College or university teachers as members of disciplinary communities of practice enjoy academic freedom against orthodoxies imposed from outside—we'll leave to the next section the important question of outside what, whether the disciplinary community of practice or the institution in which the faculty member teaches. In exchange for that freedom, they must display a reasonable fidelity to the goods and standards of the profession itself as expressed locally or nationally or both. If these goods and standards argue against even discussing the bearing of religious perspectives on what they teach, appeals to academic freedom will be of little avail if they choose to violate them.

For students this means that they are unlikely to hear much about religion's bearing on issues of study unless the discipline allows it or the faculty member is willing to buck the discipline and bear the consequences. In the chapter "Reticence," I address the question when a faculty member might think it not inappropriate to take the risk and why. For now it suffices to acknowledge that it is a risk, and one unlikely to be covered either by academic freedom as traditionally understood or by the responsibilities entailed in that freedom for the proper treatment of students.

But what of the responsibility most clearly enunciated in the 1915 *Declaration* that faculty as professionals owe their students a fair and balanced presentation of divergent opinions? Again, recall that it falls to a competent practitioner and the disciplinary community to which he or she belongs to decide (1) what constitutes a fair and balanced presentation of (2) which divergent opinions. If religious perspectives are not seen to qualify, that's that. Within the American context, the only route to change that is consistent with its traditions is for faculty themselves to become more self-aware of the bearing of religious or spiritual beliefs and practices on scholarship, teaching, and human understanding and then broach the question with colleagues. A good place to start is with self-reflection and collegial conversation.

## Institutional Academic Freedom

From the perspective of the disciplinary communities of practice, any concession of decision regarding academic freedom to the local colleges or universities can be seen as a dilution of the disciplinary communities' rights and responsibilities. The AAUP's founders disapproved strongly of educational institutions that were, in their words, "designed for the propagation of specific doctrines prescribed by those who have furnished its endowment." They were stigmatized as "proprietary schools," arguing that "[t]hey do not, at least as regards on particular subject, accept the principles of freedom of inquiry, of opinion, and of teaching; and their purpose is not to advance knowledge by the unrestricted research and unfettered discussion of impartial investigators, but rather to subsidize the promotion of the opinions held by the persons, usually not of the scholar's calling, who provide the funds for their maintenance." While they professed (somewhat disingenuously) no desire to express an opinion on the desirability of the existence of such institutions, they wanted to assure that such schools not fly under false colors and to deny them the title of true university.[42]

By 1940, and in order, no doubt, to secure the support not only of the AAUP but also the Association of American Colleges (since 1995, the Association of American Colleges and Universities), a national association made up of college and university presidents, the paradigmatic *Statement* on academic freedom in America makes provision for limits on academic freedom in church-related colleges and universities, specifying only that such limitations should be clearly spelled out contractually at the time of hire. The subsequent 1970 *Interpretive Comments* takes this provision back, asserting simply that "[m]ost church-related institutions no longer need or desire the departure from the principle of academic freedom implied in the 1940 *Statement*, and we do not now endorse such a departure."

While I think it probably true that "many" church-related institutions of higher education today no longer need or desire this "limitations" clause, to claim that "most" do is, as one authority on academic freedom puts it, "highly presumptuous and overly simplistic."[43] As Robert Poch, the associate commissioner at the South Carolina Commission on Higher Education and academic authority on academic freedom, points out, this implies that the AAUP "knows definitively whether church-related colleges and universities need a departure from academic freedom as defined by the association and refuses to consider the possibility that different constructions of 'truth' and 'ways of knowing' exist in academe." To put the second half of this objection in terms used earlier in this book, the *Interpretive Comments* suggests that the AAUP is the only community whose goods, standards, and practices count. This is at least debatable, especially given the differing goods, standards, and practices found with the various academic disciplines much less differences among academic and religious communities of practice.[44]

The *Interpretive Comment* also suggests that the AAUP definition of academic freedom should trump institutional definitions shaped by values and beliefs of sponsoring denominations. If allowed to carry the day, the AAUP *Interpretive Comment* would undermine attempts by church-related colleges, universities, or seminaries to retain their distinctive identities. For this and related reasons, some church-related colleges and universities have explicitly chosen either to adopt the 1940 *Statement* without the 1970 *Interpretive Comment* or to substitute their own declarations regarding academic freedom.[45]

Finally, as Poch points out, the AAUP approach emphasizes the academic freedom of individuals. The American courts, for their part, have tended to recognize *institutional* academic freedom. As Justice Frankfurter wrote in *Sweezy v. New Hampshire* (354 U.S. 234, 263 [1957]), academic freedom entailed "the four essential freedoms of a university—to determine for itself on academic grounds who may teach, what may be taught, how it shall be taught, and who may be admitted to study."[46] I believe that we see in this legal decision an awareness of the role of community—in this case, institutional community—in establishing and enforcing goods, standards, and practices. It allows for diversity among institutions by lodging authority on key issues of who teaches, what is taught, how it is taught, and to whom it is taught in the control of the local institution.

Whatever one thinks about the 1970 *Interpretive Comment*—and for my part, I found in it no threat to the Lutheran identity of St. Olaf College—it would be prudent for all faculty to apprise themselves of the policies of their own institution regarding academic freedom. Most colleges and universities should say explicitly what their policy is, whether it follows the

AAUP 1940 *Statement* and the 1970 *Interpretive Comments*, and if not, what departures from the AAUP's position on academic freedom may obtain.

## Religion and Institutional Academic Freedom

However one evaluates the intellectual cogency of the AAUP's drive for one standard, it should not surprise the observer that the very diversity found in American higher education regarding mission and purpose has forced accommodations. Let me briefly mention two.

First, let's begin with the happy thought that most of the time most professionals are likely to behave as professionals should. That's a primary goal of the extensive professional formation they undergo. Yet being fallible human beings, they may slip and when they do, something needs to be done to put matters right. Some entity must be able to enforce professional responsibilities when individual members of a disciplinary community of practice behave unprofessionally or are accused of behaving so. It is unrealistic to expect such procedures to work only at the level of the national disciplinary guild. As a practical matter if nothing else, a student who feels a faculty member has behaved unprofessionally needs a local venue for airing and resolving his complaint. This requires the local institution to establish policies and procedures that fit the local circumstances. Institutional academic freedom arises in part out of this practical need. It is normally a cooperative venture, where the institution through its faculty—that is, through the local representatives of national disciplinary communities—develops policy and procedures for adjudicating and rectifying complaints. Faculty (sometimes together with administrators, sometimes alone) staff the bodies that hear and decide complaints. This is professional self-regulation at the local level.

There is a second and more controversial way in which institutions exercise an academic freedom that may limit the research and teaching of individual faculty members. Recall the role of disciplinary communities of practice in determining the truth of matters as discussed in the chapter "Community Warrant." Scholars approach the (provisional) truth of matters through a disciplinary community's collective efforts at research, testing, peer review, debate, and so on. This is a determination of the truth of matters by revisable group consensus. Other communities (such as religious communities) may decide to determine the "truth of matters" differently.

In the matter of academic freedom, the community's role in establishing the truth of matters can lead to different criteria. The AAUP favors the judgment of the community of competent inquirers. But with colleges and universities founded by members of Roman Catholic orders, for example, the community will commonly look to two authorities, the magisterium that subsists in the bishops of the church and, if necessary, in the pope alone, and the community of competent inquirers' current, revisable consensus. The controlling document in current discussions regarding academic freedom at Catholic colleges and universities is *Ex corde ecclesiae* (1990). The document acknowledges both institutional autonomy and individual academic freedom "so long as the rights of the individual person and of the community are preserved within the confines of the truth and the common good."[47] For some colleges and universities that look to Reformed tradition within Christianity, the determination of the truth of matters may rest on the current, revisable consensus within a shared and covenanted Christian worldview.[48] Other arrangements are possible, but these three examples illustrate the thinking that underlies an understanding of institutional academic freedom that relies on community warrant.

In America, academic freedom arose as a by-product of the development of professional disciplinary communities. It is primarily a freedom that society has granted to faculty in exchange for services rendered in advancing knowledge. It is accorded with the understanding that the disciplinary community will responsibly police its own. Students do not possess freedom in this sense. Instead, their academic freedom arises out of the faculty's professional obligation to secure conditions conducive to the ability of students to learn. In practice, and often in the eyes of the court, faculty academic freedom depends on institutional academic freedom. And because colleges and universities may differ in their understanding of how the truth of matters is best secured, faculty academic freedom (and concomitant responsibilities) may also vary from institution to institution. Any conversation about academic freedom and religious expression must engage the interplay of contending freedoms, the role of disciplinary communities in establishing and enforcing standards, and the limitations on some freedoms that are entailed by the securing of others.

# Chapter 10

# Reticence

In the chapter "Inclinations," we explored how religious or spiritual convictions may *incline* those who are religious or spiritual to favor one explanation or interpretation over another, even as the scholars in question commonly *justify* their explanation or interpretation without recourse to their religious or spiritual beliefs. Certain questions naturally arise: Are there good reasons to be reticent about such convictions when framing a scholarly argument? Are there cases where such reticence is unnecessary or even inappropriate? What objections are likely to be raised when explicitly religious or spiritual claims are advanced? How will your disciplinary or institutional community likely respond?

Before tackling these questions, we need to remind ourselves once again that there is no one pattern that will fit each and every scholar. Some scholars are likely to be more sharply inclined by their core convictions than others. Some scholars belong to traditions, be it religious or spiritual, that expect them as scholars to draw explicit connections between their core convictions and scholarly interpretation; others do not.

Most importantly, scholars may need to adapt to the expectations and limitations imposed by their field. In the humanities, for example, religious and spiritual perspectives, where relevant, can relatively easily be voiced in scholarship or teaching because there is such widespread awareness within the humanities fields that interests and convictions influence our knowing and interpreting. As long as the scholar or teacher avoids the obvious dangers and pitfalls, she can probably (if she wants to) bring explicit but circumspect religious or spiritual convictions into her scholarship or teaching with few serious objections from colleagues.

At the other extreme is the natural sciences, with its scientific "background beliefs" about proper knowledge, inquiry, and explanation, in which religious issues and questions are rarely pertinent or mentioned, and many of you who are natural scientists automatically keep your religious or spiritual convictions (if any) separated from your scholarly work and teaching. The reflections initiated by this book may incline you to break through this separation (at least to a modest degree) and expose students to the role of religious conviction in influencing vocational choices regarding field and research interests. Further, and more demanding, these conversations may encourage you to help students understand how science has become the dominant model—the reigning ideal as it were—of what true knowledge, inquiry, and explanation should be, and to be clear with your students about the scope of this model and its implication for other, competing forms of knowledge, inquiry, and explanation.

Caught between these extremes, social scientists will likely find it more difficult than do colleagues in the natural sciences or the humanities to square the mention of explicitly religious considerations with disciplinary norms and with the expectations of fellow social scientists. For complicated reasons of history and subject matter, the social sciences and the various religious traditions compete for much of the same intellectual territory and have ample reason and occasion to clash. To the extent that the social sciences are *sciences*, they embrace the assumptions that undergird the natural sciences. But to the extent that the social sciences are *social*, they must also contend with what difference it makes that the subject of their study is human beings who, unlike flora and most other fauna, evince purpose, seek meaning, and act on values.

In short, each scholar will have to decide what is best in a particular situation. The important thing for scholars is to know when and why it is necessary *for them* to mention, or not to mention, religious or spiritual convictions in their teaching or scholarship or both.[1]

## Why Be Reticent?

Reticence about one's religious or spiritual convictions is the default mode today for most scholars in most colleges and universities. Faculty do not generally mention their personal convictions in their writings or their teaching. To lay the groundwork for conversation—with one's own self or with colleagues—let me offer some observations that have helped me think about this (often self-imposed) restraint. I speak now as an academic who is religious and who has consulted with other academics who consider themselves religious or spiritual.

1. Most commonly we say nothing because there is nothing to say. Our religious or spiritual beliefs simply have no bearing on our immediate

research or teaching. We commonly spend the bulk of our time on technical activities—for example, reading texts, doing experiments, reviewing the relevant literature, explaining the field, and so on. Our religious or spiritual convictions may urge on us the requisite self-denial, diligence, and honesty needed to do such technical activity well, but they have little or nothing to say about the content of what we're doing. We may also have been influenced by our religious beliefs and practices in our choice of field or the topics of our research and teaching (see the chapter "Narrative Identity"). But our beliefs may nonetheless not add new or different perspectives to the material we're studying or presenting.

2. Even if our religious or spiritual convictions do have a bearing on our research or teaching, they may not further our argument in any significant way. A nonreligious scholar may well reach much the same conclusion or advance much the same argument as we do. After all, one does not have to be religious or spiritually inclined to, say, oppose some forms of human experimentation, to be morally critical when recounting the history of Nazism in Weimar Germany, to raise questions about evolutionary psychology, or to find fault with Rational Choice theory. Secular scholars may take much the same interpretive or explanatory tack for reasons that have nothing to do with religious or spiritual conviction and everything to do with widely shared moral norms and (to be sure, often contested) considerations of evidence, reasonable argumentation, and demonstrated results. And, to insist on appropriate complexity in such matters, to advocate a controversial position on the use of human subjects, to attempt to understand sympathetically and nonjudgmentally the attraction of Nazism to Weimar Germans, or to see the strengths in the various theories of evolutionary psychology or rational choice do not necessarily make one antireligious. The issues may be complicated, debatable, and amenable to a variety of justifiable resolutions. In many cases, the cogency of the argument depends on the strength and soundness of the grounds, warrants, backing, and qualifications that we advance. Nothing may be added by bringing up our religious or spiritual convictions, and something may actually be lost.

3. Finally, for good and weighty reasons, we may be reluctant to bring up our religious or spiritual convictions because we're aware of the unhappy history of discrimination, violations of academic freedom, and violent resolution of disagreements occasioned by religious traditions, and in the West especially by Christianity. "Cautionary Tales" offers a brief overview of the real dangers that have attended the mixing of religious or spiritual conviction with scholarship and teaching. So we may not bring up our religious or spiritual convictions because we judge the loss in candor to be preferable to raising even the specter of a return to religious intolerance. This is not an inappropriate concern, but it entails tradeoffs that we need to ponder.

## Why Be Forthcoming?

Why, on the other hand, might it be worthwhile to advance a specifically religious or spiritual claim in one's teaching or research, even if such mention may not be necessary? Here are three reasons; you may be able to generate more.

1. Such frankness may open up opportunities to understand better our beliefs and their bearing on, and implications for, our disciplinary work. When we communicate according to rules set by others and shoehorn our beliefs into an alien formulation, we may be doing harm to ourselves, our beliefs, and, yes, even our scholarship.
2. Such frankness may also allow others to understand better and critique what we're doing. It may be more honest to give critics a crack at what we really think and why.
3. Such frankness may, for some, constitute obligatory witness.[2] Religious traditions may require of their adherents a public testimony of some beliefs or a public display of certain practices or both. Traditions vary and so do members in their degree of fidelity to their community's expectations. Recall as well that other ideological communities also encourage "going public." Not all true believers or orthodox practitioners are religious.

While I can understand why religiously or spiritually inclined scholars might be moved by such considerations, I see few occasions where it is obviously *necessary* to offer explicitly religious or spiritual claims to support a particular interpretation or explanation. As I have suggested above, much of the time scholars can justify their interpretation employing secular grounds and warrants. And they may be well advised to do so, given some of the considerations we just rehearsed.

But there are at least three areas where explicit religious or spiritual warrant may be appropriate and even required. You may disagree, and you may want to add or subtract from the following list.[3]

## About What Types of Questions?

The chapters in part II suggest that, among other things, religious traditions (and most spiritual worldviews) had a stake in how one thinks about morality, understands human nature, and construes the cosmos. These three broad topic areas do not in any way exhaust the sort of convictions that characterize religious communities of practice, but they do delineate three areas where religious or spiritual conviction obviously bears on interpretations advanced within the modern American academy and, if advanced explicitly, may enrich and deepen academic conversation.

## Moral and Ethical Claims

When scholars or teachers make moral or ethical judgments about what should be studied or about the use to which such scholarship should, or should not, be put, they may want to alert their students or readers in those cases where their moral judgment comes out of commitments to a specific religious or spiritual tradition.

For example, various religious and spiritual traditions may oppose on moral grounds embryonic human stem cell research because the stem cells are collected from (very early stage) human embryos produced in fertility clinics or developed for the explicit purpose of embryonic stem cell research. If a biologist belongs to such a community and shares its views, I see no compelling reasons why in stating his moral reservations he should not refer to the religious or spiritual source of his objection.[4]

Of course, as with any moral or ethical claim, the person advancing it will need to be prepared to argue for its validity and be willing to consider its broader implications. Moral arguments may not always converge on consensus, but in academe at least they deserve more than mere assertion and counter-assertion. In fact, given the academy's commitment to the pursuit of understanding, moral disagreements that remain unresolved can nonetheless deepen the participating parties' understanding of each other, of the contending positions, and of their own views and their implications. This is no small gain.

## Claims about Human Being

Various social sciences such as psychology and economics may base their theoretic edifice on assumptions about "human nature" and "human flourishing" that would be contested by many religious and spiritual traditions. These assumptions—for example, the model of the rational, self-interest maximizing human being that underlies many economic models, or the assumption underlying some theories of psychology that psychological health consists largely in individual self-development and self-expression—may be "givens" within their respective field (or subfield). Even so, such assumptions are neither empirical generalizations nor tested propositions. Rather, they reflect a particular set of values that may not be shared by religious or spiritual traditions.[5] When dealing with generalizations about such things as human nature, why should one set of assumptions—say, Enlightenment views of humanity and human flourishing—be privileged over religious or spiritual alternatives without an explicit case made for the preference?

As with moral or ethical claims, once the assumptions are raised up out of the background and put to the question, it may be necessary for the discussants to make explicit the commitments underlying their preferences. In such a debate, religious or spiritual commitments should be allowed to enter into the debate on equal footing with secular commitments. A religious or spiritual perspective should not be privileged over a secular view, nor should it be discounted just because it is religious or spiritual.

## Metaphysical Claims

Occasionally, scholars in the natural sciences and social sciences confuse science with metaphysics (or at least that branch of metaphysics that concerns itself with a "maximally comprehensive view of reality"[6]). In short, they draw conclusions that exceed science's grasp. For example, a biologist may confuse methodological naturalism, which assumes methodologically that scientifically adequate explanations for a natural biological phenomenon should be supplied by causes and factors that do not refer to the divine, with metaphysical naturalism that denies "that there exists or could exist any entities or events which lie, in principle, beyond the scope of scientific explanation."[7] In weighing this metaphysical (and nonempirical) claim, the scholar who is religious or spiritual may wish to point out that metaphysical naturalism is an assertion of philosophic opinion rather than a statement of fact, scientific or otherwise, and is not subject to scientific proof or disproof. To make this limited point, the scholar can simply draw attention to the unwarranted move from methodology to ontology. He or she may, however, want to go further and offer an alternative metaphysical view, one derived from religious or spiritual commitments. More on this in the next section.

Once again, the sharing of religious or spiritual commitment may be appropriate so long as the critic recognizes that the religiously or spiritually based metaphysical view has no more standing as a *scientifically valid* argument than does metaphysical naturalism. And as with moral arguments, disputes regarding one's metaphysical commitments may not be resolvable by argument, but in academe a good spirited argument should nonetheless be expected.

## What about Disciplinary Standards and Practices?

Some critics may argue that the use of religious or spiritual claims violates the standards or practices of their disciplinary community. The critic contends either that the claim fails to meet disciplinary standards or that the claimant fails to abide by disciplinary practices (or both). Regarding the claim itself, for example, a critic in the natural sciences may argue that a religious or spiritual claim lacks the empirical grounding or testable warrant expected of scientific claims. Regarding the claimant, a critic may charge that a religious or spiritual claimant simply asserts his claim as true without entering into the discipline's accepted practice of offering grounds and warrants for his claim. In other words, the claimant (so it is charged) is close-minded and deaf to argument, and thereby violates key practices constitutive of the discipline.

Let's dispose of the last point first. Some claims may well be fundamental in the sense that the claimant can't give any more basic reason and shouldn't really be pressed to do so. He simply believes, say, that lying is wrong. As ethicist Jeffrey Stout suggests, to attempt to push the respondent beyond this basic belief by asking *why* he thinks lying is wrong may be to engage in Socratic bullying.[8] In such cases it would be unfair to brand the claimant as close-minded if he refused to offer grounds or warrants for this claim.[9] Most of the time, however, it is not only fair but also appropriate to expect a claimant to advance a reasonable argument, complete with grounds, warrants, backing, and the like, for a claim that he advances. If he refuses to do so, then critics have a presumptive right to accuse him of violating the accepted standards and practices of most disciplines. I see no reason why religious or spiritual claims should be treated any differently in this regard.[10] This is especially true of claims that admit of direct or indirect empirical test. In short, more should not be expected of religious or spiritual claims than is expected of nonreligious claims—but also not less.

The conversation about disciplinary standards and religious or spiritual claims can, of course, be foreclosed before it really gets started. A critic may simply assert that the standards and practices of the discipline prevent accepting *any* religious or spiritual claims (or at least no claims that cannot be restated in secular terms).[11] Although this may be true of some disciplines, it rather begs the question of why religious or spiritual claims should be barred. It seems only fair, in the face of such blanket exclusion, to ask why religious or spiritual claims are deemed ipso facto inadmissible. Once reasons are given, discussion and argument may proceed on the merits of prohibition.

If you are able to get beyond a blanket prohibition, the next step is to focus on domains in which explicitly religious or spiritual claims are most likely to arise. I have suggested that it may not be inappropriate, and perhaps even seem necessary to some scholars, to advance explicitly religious or spiritual claims when arguing moral questions, characterizing the fundamental nature of human being, and taking positions that are rooted in maximal views of reality (i.e., making metaphysical claims). To argue about morality or metaphysics takes the scholar outside his or her disciplinary community of practice into new encompassing communities with their own goods, standards, and practices germane to such discussions or arguments. To advance fundamental claims about human nature may also entail engagement with moral or metaphysical questions. But a good deal that is asserted about human being may also give rise to legitimate inferences that can be tested empirically or subjected to rational scrutiny or both. For example, claims about human rationality that lie at the foundation of the Rational Choice theory and much of neoclassical economics yield testable results that have led scholars such as Daniel Kahneman, Amos Tversky, Donald Green, and Ian Shapiro to challenge the adequacy of the underlying assumptions.[12] The same should be true of certain religious or spiritual claims about fundamental human nature. When generalizations lead to testable hypotheses, I see no reason why the appropriate disciplinary goods, standards, and practices should not apply, whether the generalizations arise from religious or spiritual claims or from, say, interpretive schemes that arose out of the European and American Enlightenment.

## Moral Reasoning

The community of moral inquirers encompasses the various disciplinary communities and the larger society of which they are a part. In other words, moral claims draw on moral intuitions, maxims, and practices of moral reasoning shared within the larger society of which the disciplinary community is but a part. At issue, then, is not whether religious or spiritual claims are appropriate in terms of the goods, standards, and practices of the disciplinary community, but rather whether religious or spiritual claims meet the standards appropriate to the broadly encompassing community of moral inquirers in which disciplinary professionals have no particularly privileged position.[13]

Not that moral questions have nothing to do with disciplinary goods, standards, or practices. When the moral reasoning concerns disciplinary practices (e.g., the morality of certain forms of research), the disciplinary domain will provide the context and grounds on which the moral reasoning proceeds. Consider again the debates over the morality of embryonic stem cell research. A microbiologist can specify when an embryo is likely to first experience sensations, but cannot on the basis of his or her specialized knowledge specify that an embryo is (or is not) a human being with full moral status when it has achieved this stage. Others have as much right to argue this point as the biologist, and whatever view on this issue is advanced, it must be argued for keeping in mind the (contested) goods, standards, and practices of the larger community (or communities) of moral inquirers.[14]

A critic may, of course, attempt to limit a discipline's membership in this encompassing community of moral inquirers. He may assert, for example, that the generally accepted disciplinary standards distinguish between facts and values and bar certain questions of value (i.e., moral questions) in properly formed disciplinary arguments. Much could be said about this, starting with the question whether the fact–value distinction is even cogent given what we now know about situated human reasoning.[15] Be that as it may, moral disagreements tend to arise in a discipline when considering disciplinary *practices* or when expressing a judgment about the *behavior or moral beliefs of those one is studying* (say, in history or anthropology.) We've seen how questions regarding the morality of disciplinary practices move the argument into a wider community of moral inquirers with its own goods, standards, and practice. When making moral judgments about the behavior or beliefs of those we are studying, the scholar will be expected by colleagues to be sensitive to the considerations and associated literature within each discipline regarding cross-cultural understanding and judgments—a set of standards and considerations that developed over the years, in no small part, through engagement with the larger community of moral inquirers.[16] And in any case, at this point, we're discussing the cogency, soundness, and appropriateness of particular moral judgments, not whether religious or spiritual claims have any role to play. Arguments are resolved on the merits.

Some critics will also insist that in our liberal pluralistic society, which includes higher education, religious or spiritual moral claims have no place unless they can be reframed in secular terms. Many scholars have written advocating or attacking this "preclusion" in the public political arena. It is beyond the scope of my argument to rehearse all the considerations raised in this debate.[17] It may, however, be worth asking whether a (highly contested) ban against religious or spiritual claims in public political debate should be applied to the academy. Does an institution dedicated to advancing knowledge need to adopt standards different from those applicable to a society that, for its flourishing, needs to reach at least a working consensus on the common good? Put another way, to what extent is it incumbent on an institution dedicated to advancing knowledge that it tolerate and even encourage honest (even though passionate and perhaps irresolvable) intellectual disagreement?[18] Is knowledge not advanced when contending parties acquire a better understanding of each other, of their respective moral positions, and of their own positions and their implications even though no agreement is reached?[19] Are not advancements in knowledge and understanding what the academy is all about?

## Metaphysical Reasoning

In the chapter "Religious Formation," I suggested that for the purposes of discussion it was useful to adopt George Lindbeck's short characterization of religions as "comprehensive interpretive schemes, usually embodied in myths or narratives and heavily ritualized, which structure human experience and understanding of self and world."[20] If we bracket the bit about myths, narratives, and ritual,[21] this also describes metaphysical worldviews. They are (maximally) comprehensive interpretive schemes that structure human experience and understanding of self and world. And like religious worldviews, the well-tested ones tend to defy either proof or refutation. In fact, they are often so encompassing and powerful that their holder takes them as self-evident. She may then find it difficult even to recognize that alternative construals may be for their adherents as compelling and commonsensical as she finds hers to be. Arguments may not get very far in such cases.

In today's academy, the dominant secular alternative to religious comprehensive interpretive schemes tends to be some variation on naturalism.[22] Naturalism is the overwhelmingly favored comprehensive interpretive scheme in no small part because of its remarkable successes in answering questions in the natural sciences and (to a much lesser extent) in the social sciences. Given this dominance, it may be helpful to point our briefly where the metaphysical clash between variants of naturalism and different religious or spiritual comprehensive schemes tends to occur.

"Naturalism" means, and has meant, a number of different things to philosophers, theorists, and other scholars.[23] Here's one short definition from the *Oxford Dictionary of Philosophy* that captures the major lineaments of this view (while necessarily passing over the many nuances and exceptions that any fuller account would require):

> **naturalism**. Most generally, a sympathy with the view that ultimately nothing resists explanation by the methods characteristic of the natural sciences. A naturalist will be opposed, for example, to mind–body dualism, since it leaves the mental side of things outside the explanatory grasp of biology or physics; opposed to acceptance of numbers or concepts as real but non-physical

> denizens of the world; and opposed to accepting real moral duties and rights as absolute and self-standing facets of the natural order.[24]

As this definition illustrates, naturalism is defined as much in what it tends to oppose as to propose. It is akin to, but not necessarily identical with, materialism or physicalism—the view that the world is entirely composed of matter, or that the world contains nothing but matter and energy and that entities have only physical properties. But what it insists on, as philosopher Alan Lacey observes, "is that the world of nature should form a single sphere without incursions from outside by souls or spirits, divine or human, and without having to accommodate strange entities like non-natural values or substantive abstract universals."[25]

This short definition suggests where disagreements are likely to pass over into the realm of clashing comprehensive interpretive schemes. Here are three prime candidates likely to arise in academic conversations[26]:

**Values, norms, and moral claims** In what sense are values, norms, and moral claims real? In what sense can they be said to be true or false? subjective or objective? In what sense are they even meaningful? And if so, what do they mean? How does one argue for or against a value, norm, or moral claim? and with what grounds and warrants?

**Minds, mental phenomena, and souls** In what sense can minds or consciousness be said to exist? Is there such a thing as a soul, and what is its relationship to the body? In what way do minds or mental phenomena depend on the body and its physical processes? Are they determined by the underlying physical processes? If so, with what implications?

**Free will, responsibility, and determinism** We believe we make free choices, but do we? In what ways are we determined by our underlying biology? by natural, lawful causality? by our evolutionary, genetic heritage? How do we adequately explain human behavior? according to what models and procedures? If free will is an illusion, should we be held responsible for our choices? Is the cosmos itself determined? in what ways? How does one account for creativity or spontaneity or apparent contingency?

When differences of opinion surface in these and related areas, it is worth considering whether deep metaphysical differences—differences in comprehensive interpretive schemes—are at the root. And if they are, *according to which comprehensive interpretive scheme(s) are the subsequent arguments to be judged*? Before the specific, occasioning claim can be addressed, discussants must agree on the ground rules as to what will be taken to be acceptable grounds and warrants. This moves the debate beyond the immediate goods and standards of the disciplinary community of practice into a realm where comprehensive interpretive schemes contend with each other.

A critic may at this point argue, and with considerable historical justification, that in modern American higher education, the naturalistic comprehensive scheme *is* the accepted scheme according to which such disputes must be resolved. As mentioned earlier, such an assertion, while having past practice on its side, rather begs the question *why* religious or spiritual alternatives should continue to be banned, especially given today's highly diverse and pluralistic academy. If the objector is willing to engage this question, discussion and argument may proceed on the relative merits and demerits of the contending schemes and their proper role in higher education.

To be realistic we need to recognize that it is very difficult to convince anyone that his or her deeply held metaphysical comprehensive scheme is so wrong headed that he or she should abandon it in favor of another. Of course, it is not impossible. Minds can sometimes be changed. But it commonly requires developing at least the initial arguments for change *from within* the comprehensive interpretive scheme that one seeks to critique. Only after internal discrepancies and tensions have surfaced and are acknowledged as telling, is the adherent likely to be open to alternative construals.[27] This is, however, no easy task.

The bottom line for our purposes: metaphysical claims in disciplinary discourse move the conversation into a new domain where the contending schemes even define what is considered to be an acceptable, much less a telling, argument. I see no reason why religious or spiritual schemes should ipso facto be excluded from this contest, although we all need to recognize that in such contests the first argument is going to be about the standards for argument and not about the metaphysical claim that sets the argument going in the first place.

* * *

Philosophers and other theorists have spilled barrels of ink disagreeing on how one properly reasons about morality, ethics, human nature, metaphysics, and so on. If you have colleagues who have read this literature, you may want to ask them for an overview. I suspect, however, that a discussion of this literature and the issues it raises will take more time than you can spare and will lead to little, if any, consensus among conversationalists. But I do think there is a defensible way of evaluating these and related disciplinary standards *in practice*, and that is to ask the question of consistency.

## The Argument from Consistency

As we've just briefly seen, religious or spiritual traditions are not the only comprehensive interpretive schemes, especially in the academy. Various forms of naturalism offer comprehensive ways to structure our understanding of self and world. With perhaps less relevance to cosmology, Marxism, feminism, and various schools within philosophy and the social sciences also offer comprehensive interpretive schemes that structure how their adherents understand self and at least the social world.

Given these similarities (and differences), can these wide-ranging but nonreligious interpretive schemes be distinguished from religious or spiritual schemes in a way that exempts them from the objections considered earlier? What is it about religious or spiritual views on morality, human being, or metaphysics that so differ from nonreligious analog that it disqualifies the explicitly religious or spiritual from academic discourse?

Take the three broad intellectual areas in which (I suggested above) explicitly religious or spiritual claims may be appropriate or even required if one is to be intellectually honest—that is, moral claims, claims about human being, and claims that are rooted in contending maximally comprehensive views of reality. What would colleagues from different disciplines say to differentiate nonreligious claims at home within their discipline from religious or spiritual claims?

In addressing these comparisons, it is crucial to recognize *variety* within such interpretive schemes. On most contested issues in the academy regarding morality, human being, and metaphysics, you'll find considerable diversity of opinion among Christians, Jews, or, for that matter, adherents to the naturalistic paradigm, feminists, or Marxists. There is no one Christian position on the moral status of the embryo, for example. Rather there are various liberal views, differing evangelical views, a variety of Catholic positions, and so on. Similarly, there is no one feminist view of human being, rather liberal feminists may disagree with radical feminists, psychoanalytic feminists with postmodern feminists. Even such labels may disguise significant differences among Christians or adherents to the naturalistic paradigm, Marxists, or feminists. There is comparable variety among scholars who call themselves Jewish or Muslim, Rational Choice theorists or social behaviorists, and so on and so forth.

Although I don't want to prejudge the range of views represented in your own institution, my experience suggests that you'll almost certainly be able to surface nonreligious cases that fail to meet even a reasonably lax interpretation of the standards that are used to bar religious or spiritual claims. So why the seeming inconsistency? Here's one possible answer: Despite analogous shortcomings found in nonreligious interpretive schemes, the permitted violators commonly lack the long history of imposed orthodoxy and interference in academic affairs that characterizes the (Christian) religious tradition in Europe and America.[28] An unhappy history is certainly not the whole explanation, but the cliché may still fit the situation: Once burned, twice shy.

## What about the Separation of Church and State?

A few words should probably be said about a charge most often heard in secular universities, that the mention of personal religious or spiritual conviction in scholarship or teaching violates the constitutional separation of church and state. This objection seems to be based on several serious misunderstandings. To summarize briefly what could be an extended argument,[29] this First Amendment argument presents as a stark separation what is in fact a complicated balancing act even within our political system (which is *not*, in any case, academe). Further, it overgeneralizes the prohibitions of the establishment clause of the First Amendment at the expense of the rights of free exercise enshrined in the exercise clause. Finally, and this is crucial to the question at hand, it draws an unwarranted analogy between the academy and the public political arena. So long as the scholar and teacher avoids blatant proselytization and remains sensitive to the unequal power relations between faculty and students (or between senior faculty and junior or probationary faculty), the explicit mention of religious or spiritual conviction in scholarship or teaching may remain problematic for other reasons, but not because the First Amendment requires such beliefs to remain private.

## What about Collegial Condescension?

Whatever the field, one will always find colleagues who will give you a hard time. This happens to scholars who are feminists and traditionalists, Marxists and Straussians, modernists and postmodernists—and religious or spiritual.

At every institution (but perhaps especially in research universities and on the two coasts) we can find academics and intellectuals who treat religious or spiritual convictions dismissively, as a "conversation stopper," as something that true intellectuals no longer take seriously. In making such dismissive judgments, they often show little familiarity with today's lived religion or spirituality. (To be fair, religious folks often treat secular intellectuals dismissively, assuming, it seems, that if these intellectuals are opposed to religion, they have nothing useful or worthwhile to say.[30])

Academics may feel inhibited about discussing the proper role of religious or spiritual convictions in scholarship because colleagues often assume nonchalantly that no true intellectual could be religious, and especially not traditionally religious. For example, Richard Rorty in his review of Stephen L. Carter's *The Culture of Disbelief: How American Law and Politics Trivialize Religious Devotion*[31] suggests that the "big change in the outlook of the intellectuals—as opposed to a change in human nature—that happened around 1910 was that they began to be confident that human beings had only bodies, and no souls." Because it is difficult, Rorty asserts, to disentangle a belief in an immortal soul from the belief that the soul is sullied by sexual acts,

> the biggest gap between the typical intellectual and the typical nonintellectual is that the former does not use "impurity" as a moral term, and does not find religion what James called "live, forced and momentous option." She thinks of religion as, at its best, Whitehead's "what we do with our solitude," rather than something people do together in churches.[32]

Leaving aside questions of the historical accuracy of his generalization, Rorty's casual assumption about the immiscible nature of true intellectual life and religious community can discourage mention of religious or spiritual convictions even as sources of insight, whether in teaching or scholarship.

The Berkeley philosopher John Searle illustrates a similarly dismissive attitude, one that seemingly frees him from confronting the possibility that what he understands as religion has at best a faint resemblance to the lived religion of religious colleagues. For example, after sketching as his view of "ultimate reality," Searle confronts the question "what about God?" If God exists, certainly God is the ultimate reality. Searle continues,

> [i]n earlier generations, books like this one would have had to contain either an atheistic attack on or a theistic defense of traditional religion. Or at the very least, the author would have to declare a judicious agnosticism. . . . Nowadays nobody bothers, and it is considered in slightly bad taste to even raise the question of God's existence. Matters of religion are like matters of sexual preference: they are not to be discussed in public, and even the abstract questions are discussed only by bores.[33]

Whatever one might say about this pronouncement, there seems to be little doubt that if departmental colleagues exhibit a similarly dismissive or condescending understanding of religious or spiritual beliefs, other colleagues might understandably be reluctant to display bad taste or be the bore by mentioning such conviction.

The sociologist Peter Berger commented that a look at history or at the contemporary world reveals that it is not religion, even fundamentalist religion, that is rare but rather the knowledge of religion including fundamentalist religion. "The difficult-to-understand phenomenon is not Iranian mullahs but American university professors—it might be worth a multi-million-dollar project to try to explain that!"[34] And he concludes that "[s]trongly felt religion has always been around; what needs explanation is its absence rather than its presence. Modern secularity is a much more puzzling phenomenon than all these religious explosions—if you will, the University of Chicago is a more interesting topic for the sociology of religion than the Islamic schools of Qom."[35]

The examples from Rorty and Searle on one side, and Berger on the other, illustrate that a dismissive attitude is indulged on both sides of the divide, and those who decide to be explicit and self-reflective about religious or spiritual convictions and their influence on scholarship or teaching will need to be prepared for flak from colleagues. You need to count the costs before taking the plunge.

# Chapter 11

# In the Classroom

In the chapter "Reticence" we considered why teachers and scholars might want to be circumspect about advancing explicitly religious or spiritual considerations in conversations with colleagues or in scholarly publications. We risk the charge of irrelevance or of airing private matters if we share biographical influences; we may violate disciplinary standards if we advance explicitly religious claims or warrants. But at least in a conversation with colleagues, the give-and-take on the aptness of the religious reference and the resulting criticism have some chance of proceeding on the merits. If we limit our comments to questions of morality, metaphysics, or human being, we have some chance of securing a hearing, perhaps even a change of mind. The objector will, of course, be defending the academy's "default" position that a faculty member's explicitly religious or spiritual perspectives have no place in scholarship or teaching. This will give the objector a considerable advantage in the exchange. But we are also a disciplinary professional with authority of our own and an earned right to be heard. The "default position" can be successfully challenged if we are reasonable and responsible in making our case.

With students other considerations come into play. As we discussed in "Academic Freedom," we faculty have an unequal relationship with our students. As long as we do not blatantly violate professional proprieties, we enjoy broad authority in the classroom over what is taken to be sound opinion within our discipline. We are, of course, ultimately answerable to our disciplinary peers for our scholarly opinions. But unlike published scholarship that usually undergoes peer review, teaching after the probationary years undergoes little scrutiny so long as we don't "go around the bend." And even then, as I can testify as a former college president, there is often little that distressed peers or administrators can do about poor teaching or inappropriate comments made in the classroom by tenured colleagues.

The unequal power relations and the occasional, but sometimes dramatic, intrusion of irrelevant religious or political issues into classroom teaching allow the group Students for Academic Freedom to argue, with some color of legitimacy, that an "academic bill of rights" is needed to protect students from indoctrination or proselytization. We discussed earlier problems with this argument within the American context. For our purposes now, we need only cite this activist movement as an illustration of why faculty members need, as a matter of prudence if nothing else, to be circumspect and thoughtful when introducing material that our discipline may consider beyond its purview or even a transgression of its standards and practices.

But if an individual teacher is willing to risk mention of religious or spiritual conviction with a bearing on issues in discussion—and keep in mind that it always entails risk, given the nature of religious and spiritual commitment (or its lack) among faculty, students, and the larger public—how might this best be thought through? In this concluding chapter, I consider two strategies for the circumspect (re)introduction of religious considerations into teaching in classes where religious belief and practice is not the obvious subject matter in the class. We look, first, at faculty self-disclosure, and, next, at what philosopher Warren Nord has called "natural inclusion."[1]

## Self-Disclosure

In introducing a discussion on public policy regarding welfare, a political science professor might jokingly confide that attending Catholic Mass is her secret vice, and so, not surprisingly, her Catholicism informs her views on appropriate policy. This self-disclosure invites students to examine how their own deep commitments may influence their own thinking about the appropriate and inappropriate ways of dealing with poverty in America. If done carefully, it does not impose the professor's views on the students; rather, it helps the students understand better why the professor takes the stance she does. It also invites students to reflect on their own commitments and how these commitments influence their views on the issues in question. The successful deployment of self-disclosure of this sort requires, of course, a deft hand, perhaps a bit of disarming humor, and, most importantly of all, a willingness to entertain and encourage alternative perspectives.

Self-disclosure may serve a pedagogical purpose if religious perspectives are germane to the subject matter at hand. And how does one determine what is germane? In "Reticence" I suggested that religious traditions (and most spiritual worldviews) have a stake in how one thinks morally, understands human nature, and construes the cosmos, and something to contribute to the ongoing conversation in each of these three domains. Religious traditions may have a stake in other domains as well, but these three are the most obvious and pertinent.

Of these three domains, the pedagogical strategy of self-disclosure best serves disciplines that engage moral or broadly normative questions (i.e., questions of what is right or wrong, more or less appropriate, moral or immoral). It would be difficult to avoid engaging moral judgment when teaching a political science class on, say, "Poverty and Social Policy" that examines the causes and consequences of poverty in the United States, Mexico, and China, and explores strategies for addressing it. To be sure, the instructor may choose to frame the normative choices in secular terms (say, those prescribed by Rational Choice theory), and insist that students do the same. Or the instructor may decide, as in the example above, to invite explicitly religious perspectives into the discussion. It is the instructor's pedagogical decision. If criticized for inviting religious perspectives on the issues, she may find the considerations and arguments regarding moral reasoning I sketched in "Reticence" useful in her defense.

Many disciplines in the modern academy deal with subjects that have normative and often moral dimensions. This is especially true of the humanities and social sciences. If the instructor wishes to develop her students' capacity to reason about such matters, she may need to give permission for students to bring their religiously informed moral intuitions into the discussion. To be sure, some political scientists (to return to our illustration) have attempted to avoid explicitly moral questions by reframing the issues in terms of Rational Choice theory (or, to give a few other labels, public choice theory, social choice theory, game theory, rational actor models, or positive political economy).[2] Not surprisingly, the assumptions underlying this broad approach have moral dimensions and consequences. Many religious traditions are likely to object to the approach's heavy reliance on rationality, its standards regarding utility, and its methodological individualism. A political science professor who teaches Rational Choice theory may choose to ignore this critique—and thereby give the impression that moral questions can be elided—or she may choose to alert the students to some of the religiously contest issues.[3] We take up the latter strategy in the next section on "natural inclusion."

It is more difficult to see what purpose self-disclosure would achieve in, say, a chemistry class. Perhaps, it might relieve some anxieties experienced by more religiously conservative students that their budding interest in science need not necessarily compromise their religious convictions. But this objective, though still worthy of an engaged teacher, does not serve the subject matter in the same direct way as in the case of the Mass-attending political scientist.

In general, self-disclose *by itself* will not appreciably advance the pedagogical task when the subject matter involves contested metaphysical claims or contending claims about human nature. These issues need explaining, often in detail. If the instructor wishes to engage students on such issues, the appropriate pedagogical strategy is some variant on natural inclusion, which we take up in the next section.

Much more could be said about self-disclosure as a pedagogical strategy, but I shall content myself with raising a few cautions.

A faculty member may employ self-disclosure both to put students on notice about where the faculty member "is coming from" and to encourage students to reflect on where they "are coming from" and how it may affect their evaluation of the matters under discussion. If the pedagogical strategy works, of course, the students will share their religious (or analogous) perspectives in turn. Student contributions may be as well considered and circumspect as the faculty member's. But given their age and inexperience, students are also capable of saying things that may take the discussion in problematic directions. In our hypothetical political science class on Poverty and Social Policy, a student may respond to a faculty member's cautious self-disclosure with religious claims that may, if the faculty member is not skilled in leading discussions, derail the discussion or lead to an exchange that some students may see as disrespectful either to others or to the student venturing the remark. Consider some real-life examples: "Poverty is God's stick to get you off your butt. Welfare takes away the incentive to work and is immoral." "The case workers I interviewed for my paper were really aggressive. You know, New York Jews." "If you distribute condoms to high school students, you're just encouraging casual sex." "Abortion is murder, and it also lets girls get away with sleeping around." An experienced teacher will be able to cite many such challenging examples.

As faculty we learn how to deal with provocative student comments. There is something, however, about religious claims that distinguishes them from other controversial issues within the academy. They seem harder to argue about. They also seem, somehow, out of place. Is it perhaps the violation of our (or our discipline's) expectation that religious convictions are properly a matter of private belief, not public profession? That they are (allegedly) irrational and subjective, not amenable to rational discussion or resolution? Is our deep disciplinary formation at the root of our uneasiness?

A faculty member's self-disclosure may also invite witnessing and attempts at conversion (or attempts to correct someone who is thought to be in mortal error). What does a faculty member do if following her deliberately light comment about attending Mass as her secret vice, a student feels obliged to tell her that unless she accepts Christ as her personal savior and is born again, she's going to hell? Recent research on the attitudes of teenagers and entering students suggests that it is a rare student who is sufficiently religiously earnest and stout-hearted to attempt to lead a faculty member out of error in such fashion.[4] But it can happen, and the faculty must be capable of handling such approaches delicately. After all, if we self-disclose, we have broached the topic of religion. It would be unfair to put down a student for accepting our invitation, however ill-advised his or her comments may be.

Self-disclosure can sometimes inhibit pedagogical goals rather than advance them. We need to be cautious that our self-disclosure is not taken to be coercive or an attempt at proselytizing. Whatever our sincere intentions and pedagogical goals when we reveal our deep convictions or religious affiliation, most institutions operate with a strong presumption that religious proselytization is not appropriate on the job. We don't want to be misunderstood.

Finally, we should recognize that some students may find any mention of religious conviction, whether by faculty members or by fellow students in response, to be inappropriate or intimidating. This is especially true of students who belong to religious or secular traditions that are minorities within the college or university or within the larger society. Faculty self-disclosure may invite some students into the conversation while excluding or silencing others. Again, it will take a deft pedagogical hand to carry off the strategy successfully.

We move now from the pedagogical strategy of self-disclosure in courses that engage subjects with a clear moral dimension to another, more elaborate pedagogical proposal fraught with its own difficulties, namely, the strategy of "natural inclusion."

## Broad Natural Inclusion

By natural inclusion, philosopher Warren Nord and other commentators mean that "courses . . . that deal with religiously contested issues should at least acknowledge the existence of the religious alternatives and engage them in conversation."[5] In practice, "natural inclusion" might mean that a course in, say, economics would acknowledge that religious views influence economic activity by influencing thought on justice, human rights, suffering, the good life, and so on. It would place in conversation the different assumptions that various religious traditions and academic economics make about human nature or morality. Students should be introduced to the important controversial issues and given opportunity to explore the assumptions underlying economics and alternative views. "Put more generally," the Nord explains,

> liberal education texts and teachers should be governed by the Principle of Philosophical Location and Weight: that is, they are obligated to locate their positions philosophically on the map of alternatives, indicating what weight their views carry in the discipline and in the larger culture. If students are to be educated, they must have some sense of when they are being taught what is controversial (and for whom) and when they are learning consensus views.[6]

Such exposure, Nord argues, would be particularly appropriate in introductory courses.[7]

At this juncture, the implications may still seem rather vague and general. So let's offer a more concrete example. Take, for example, the notion of "valid" and "invalid" costs in economics. As economist Robert Nelson explains,

> In economic practice, a "valid" cost is one that consumes resources that could be devoted to the advance of material progress on earth. In order to maximize the rate of economic progress, resources must be allocated with maximal efficiency, requiring that the "cost" of every such valid item be carefully measured, helping to ensure that a great "opportunity cost" is not thereby lost. With "invalid" costs, they are defined by the fact that they have just the opposite character. To give any weight to the psychic pain or other stresses of transition and dislocation, as the economy moves from lower to higher stages of economic productivity, is to stand in the way of economic progress. Legitimate costs for economists, in short, are those that consume actual resources that can be devoted to advancing the material productivity of society; illegitimate costs are those whose introduction into the economic calculus would stand in the way of economic progress.[8]

When an economics professor reaches this economic concept of costs, natural inclusion would suggest that she acknowledge that the major religious traditions would offer a concept of "progress" different from that implicit in this economic analysis and would likely draw the line elsewhere dividing "valid" and "invalid" costs. She should, at least briefly, describe alternate views of progress and say something about economic "morality"—that which economics applauds versus that which it either ignores or deplores—and contrast it with various religious evaluations of the same set of considerations. Her discussion need not be lengthy, but it should be sufficiently detailed to alert students to the contested nature of these key concepts in economics and give them some introductory sense of the religious alternatives. This is natural inclusion in action.

Here are some further examples of religiously significant issues that one person or another might suggest could be covered in an introductory disciplinary course. Some are unlikely to raise eyebrows; others would rightly generate fierce controversy:

*Art:* the propriety of images in different religious traditions and denominations (Eastern Orthodox versus Protestant; Islam versus Christianity);
*Biology:* views of evolution, from "young-earth creationism" to the "selfish gene" theory of Richard Dawkins; views on when human life begins;
*Cosmology:* views on creation and the "big bang"; the Anthropic Principle; complexity theory and emergence;
*Computer Science:* artificial intelligence and human intelligence; views of the human mind/soul as a form of software that might be uploaded into a silicon computer;
*History:* the role of providence, chance, and free will; dispensationalism; notions of progress;
*Literature:* interpretive strategies that challenge "meta-narratives"; Theories of artistic expression that entail views of human nature, purpose, and destiny;
*Philosophy:* religious perspectives on major philosophical issues such as the nature of truth and beauty, epistemology, human nature, and ethics;
*Political Science:* views on the separation of church and state; international conflict between democratic capitalism and indigenous religious traditions and cultures; religious difference as a potential fault line in future conflict[9]; Jihad versus McWorld.[10]

This list does not exhaust the possibilities but should give a sense of the range of issues that natural inclusion, without significant restriction, might raise.

## Circumspect Natural Inclusion

Nord advocates a rather broad application of "natural inclusion." I suspect that faculty who might be willing to risk the explicit mention of religious or spiritual conviction in disciplinary courses outside religious studies will nonetheless prefer a more limited, circumspect approach for several telling reasons.[11]

1. Not all religiously contested issues deserve natural inclusion.
2. Some religious issues may deserve inclusion even though the matter is not contested.

3. Some issues, for lack of academically or intellectually credible analysis on the part of various religious traditions, do not deserve treatment unless or until this lack is overcome.

Here are some criteria that may help separate appropriate (but not necessarily necessary) candidates for natural inclusion from the inappropriate (or not yet appropriate).

*Importance:* The issue must be sufficiently pertinent to warrant the time it would take to deal with it adequately. The four following criteria factor into the determination of importance.

*Contribution:* Natural inclusion should both enrich the disciplinary analysis and also do justice to religious concerns. If the religious perspective does not deepen insight into the discipline, its methodology, assumptions, and understanding of its subject matter, faculty may in deference to the discipline's goods, standards, and practices reasonably balk at its inclusion.

*Relationship:* If there is a historic relationship between a religious tradition and the discipline's treatment of an issue, mentioning the religious connection may deepen students' understanding of it. For example, the social sciences in Europe and America grew out of (or in reaction to) Christian moral philosophy in the course of the nineteenth and early twentieth centuries and inherited questions, assumptions, and goals from this religious matrix; an explicit acknowledgment of origins and divergences may enrich the students' understanding of the material and its complexity.

*Quality:* Candidates for natural inclusion must be cogent and academically sound, a credible contribution to the scholarly search for better understanding of the subject matter. There should be readings and other resources available on the topic, written by articulate, intellectually and academically credible scholars. The faculty member raising the conversation in her disciplinary course should not have to construct the argument and analysis on her own.

*Pedagogy:* The topics for natural inclusion should pass certain pedagogical tests. They should not, for example, require of the students an intellectual or disciplinary sophistication that exceeds their educational level. Another pedagogical test is that natural inclusion should deepen the students' understanding of the material and not close off learning by reinforcing prejudices.

Before looking at the downside of all this, let's elaborate a bit on these criteria.

Religious or spiritual perspectives should *not* be naturally included in a disciplinary course on the sole ground that some topic is contested by a religious community. These is no point in insisting on the natural inclusion of positions that will only embarrass most religious people and alienate scholars, including religiously committed scholars. For example, young-earth creationism (YEC) contests biological evolution, but this is insufficient reason to include it in a biology course on evolution, except, perhaps, as an example of bad science (and possibly bad religion). Almost all biologists, including religiously committed biologists, reject YEC's claim to be science.[12] To include it in a biology course violates the criteria of contribution, quality, and pedagogy. To be sure, the claims of YEC may be important in the sense that its advocates manage to stir up considerable controversy among school boards and in the press, but this is an importance that suggests treatment in, say, a political science course, not a biology course. YEC probably even fails the criterion of relationship, since in its scientistic form it is a recent development with a spotty pedigree connecting it to earlier religious (and specifically Christian) views of Genesis.[13]

Some religious perspectives on evolution *do* meet the criteria of contribution, quality, pedagogy, and importance. To continue our example from Biology, the field of evolutionary biology itself is divided on issues ranging from punctuated equilibrium[14] to emergence, from genetic "pre-encoding" to "constructive interactionism" between genes and environmental influences.[15] Religious perspectives and interpretations may illuminate some of these issues, although ultimately the soundest science should decide the case. And it is the disciplinary professional—the teacher—who should make the call.

A plausible case could be made to limit natural inclusion to disputes internal to a discipline that offer opportunities for deepening a student's understanding of the discipline itself and its wider implications. Faculty within the discipline will need to decide what to include, and what not. Outsiders may offer suggestions, but the experts who will do the "natural including" must (and, in any case, will) have the final say.

Each community of practice, be it a religious community or an academic discipline, reveals what it values or considers significant by what it attends to and what it ignores. If an academic discipline ignores religious perspectives when, on the face of it, religious perspectives have a significant bearing on the matter under study, the discipline is making an interpretive judgment that reveals something about the conceptual scheme that frames the discipline. It may also be revealing something about the disciplinary community's heritage, tradition, and values.

When we faculty members call our students' attention to this omission, we open space to discuss the (often tacit) whys and wherefores of the discipline itself. We locate the discipline's interpretive approach within a range of alternatives. We invite our students to consider whether questions of meaning, purpose, and value have a role to play on such topics and if so, what role. We prepare our students to understand better both the community of practice that is our academic discipline and other communities of practice, especially religious communities of practice, that may interpret our subject matter differently.

The criterion of quality lies at the heart of credible natural inclusion. Much is at stake if natural inclusion is done badly, so excellence should always trump comprehensiveness. If natural inclusion extends to shoddy thinking, no one benefits. Faculty and students will be alienated, and religious or spiritual interpretations made more susceptible to easy dismissal. It is both tactically prudent and intellectually honest to include only the strongest topics (and pragmatics may suggest even greater constraints, given the limited time at a class's disposal and the many disciplinary topics that need to be covered). It should go without saying (but may not) that only faculty in the discipline can reliably identify those topics in which a religious perspective has integrity, intellectual cogency, and the supporting literature to allow a natural inclusion that both furthers the educational goals of the discipline and equips the student to understand better how various religious traditions view the life of the mind and life in the world.

The criterion of pedagogy aims to insure that naturally included material be accessible to students at the intellectual and academic level at which they are currently operating. If the religious perspective on a cosmological issue requires understanding of quantum physics, for example, there is no point in raising it in first-year mechanics. If students need to understand the nuances of (post)modern hermeneutics to follow the religious or spiritual analysis, perhaps the material should be reserved for an upper-level class or simply given over to colleagues in Religious Studies.[16] And in line with the criteria of importance and pedagogy, the treatment of religious or spiritual interpretations of a particular issue must be sufficiently pertinent educationally to justify the time it takes out of a short semester.

The criterion of pedagogy is also meant to cover unintended consequences. For example, natural inclusion might suggest that an introductory statistics course acknowledge, first, that many important things, including religious values, may not be quantifiable, and, second, that the dominance of a quantitative approach in economics and political science may tacitly undergird many injustices in modern society. Although such observations may be faithful to natural inclusion and worth exploring in detail, perhaps in a separate course,[17] raising the point at any length in an introductory statistics course may largely have the consequence of closing minds already put off by the demands of serious statistical analysis. Natural inclusion can in some circumstances become an unintended excuse to dismiss without first understanding. While some may argue that it is only fair to give "equal time" to (religious or spiritual) arguments against quantification; the pedagogical result may be an unfortunate reinforcing of math phobia. Again, the disciplinary professional needs to make the call.

## Objections

Let's assume at the outset that natural inclusion is a matter of individual faculty choice. To be sure, at some church-related schools, there may be an institutional expectation that all courses be taught, say, from a "Christian perspective" (as variously defined by the Christian denomination in question). But even in these circumstances, the institution as a whole and the department in question should nonetheless be leaving the pedagogical choices to the individual faculty member responsible for the class. (And even with this more constrained allowance for academic freedom in church-related schools, the department and institution may well still owe the larger academic and disciplinary community some justification for its imposition of this expectation.[18])

The objections explored in the chapter "Reticence" would obviously apply with special force to the practice of natural inclusion, even if the practice is a matter of individual faculty choice. To recapitulate briefly, it is widely feared that the explicit mention of religious or spiritual considerations may (among other evils)

1. justify or encourage discrimination;
2. go against disciplinary standards and practices; and
3. give rise to passionate disagreements and disputes that can lead to rancor, condescension, and even coercion or other forms of violence.

There is ample history to justify these fears. And there are further problems as well.

### Which Traditions to Include?

The world today is characterized by considerable religious and spiritual variety, and much of that variety is now at home within the United States.[19] Before teachers who wish to practice natural inclusion can decide which religiously contested issues to acknowledge in their disciplinary courses (with all the principled and practical tradeoffs discussed earlier), they must first decide which religious or spiritual traditions should have a say in deciding what's a "religiously contested" or "religiously significant" issue. For example, some varieties of evangelical Christianity might wish the professor of an introductory biology course to acknowledge the (from their perspective) contested nature of evolutionary theory. Catholic Christians might prefer that the professor acknowledge the morally contested status of, say, embryonic stem cell research. Hindus may think it most important to acknowledge the religiously problematic character of dissection and animal experimentation. Other traditions may have their own differing concerns. If all these contested issues are included, an introductory biology course will rapidly qualify for cross-listing in Religious Studies and students will find themselves shortchanged when it comes to their education in biology.

Uninformed choices are simply likely to reflect the commitments (or, more harshly, the prejudices) of the individual faculty member. Informed choice requires the faculty member to become informed, not a simple matter (see the next section). And even informed choices mean that only some limited number of religiously contested or important issues can be acknowledged in any depth and all others passed over, perhaps even without mention.

Of course we professors always chose what to cover and what to omit. But natural inclusion adds a significant new dimension to this standard pedagogical dilemma. The choice of what to include and what to exclude in, say, a biology or psychology course is likely to generate more heat among students, colleagues, and outside groups when the topic is religious belief or practice than when the topic is, say, chemistry or sociology. Any inclusion opens the teacher to the challenge: why this and not that? It is worth remembering that one of the major reasons that scholars in the late nineteenth and early twentieth centuries dropped religious (and later moral) considerations from their subject matter was to avoid just this sort of seemingly irresolvable debate.[20]

## Questions of Knowledge and Competence

To be religiously or spiritually inclined does not automatically make one an expert on religious or spiritual perspectives. In fact, some see it as a handicap that must be overcome.[21] Be that as it may, for natural inclusion to work responsibly, its practitioners need to acquire sufficient understanding and expertise that they can responsibly, in Nord's formulation, "acknowledge the existence of the religious alternatives and engage them in conversation." This will require significant investments of faculty time and probably explicit instruction or at least direction from colleagues in the field of Religious Studies.

Specialists can be skeptical regarding disciplinary interlopers, suspicious of attempts to blend methodologies and interpretive strategies, and inclined to insist that responsible interdisciplinary work requires extensive professional formation in each discipline that goes into the interdisciplinary mix.[22] This is all understandable. Our disciplinary communities insist that competence in a field requires lengthy professional formation. If we stray outside our competence, we open ourselves to charges of dilettantism, vulgarization, and lack of professional rigor.

These considerations regarding specialization and disciplinary boundaries rightly incline us to be wary of including material or perspectives that lie well beyond our area of expertise or that are claimed by specialists within another discipline. These assumptions and the resulting wariness may stand in the way of acknowledging the bearing of religious or spiritual perspectives, motivations, or concerns on disciplinary scholarship and teaching, either because religion is "outside" our specialization or because religion "belongs" to another discipline (namely, Religious Studies).

In the specific case of natural inclusion, this wariness may be amply justified. Most religiously or spiritually inclined scholars, as religiously or spiritually inclined adults generally, know only our own tradition (if we have one) and even it rather unreflectively. Given the demands of our disciplinary work, it may be unrealistic to expect us to devote sufficient time and energy to gain such expertise in the various relevant religious traditions that we could responsibly include religious perspectives in their courses. And without such study, scholars practicing natural inclusion run the substantial risk of making pronouncements about matters we understand only superficially.

When I started this project, I was attracted to natural inclusion and admired (and continue to admire) the broad and detailed argument that philosopher Warren Nord developed to justify its adoption.[23] But as I interviewed faculty around the country and as I thought through the ramifications of natural inclusion for myself, I came reluctantly to conclude that on most campuses the pursuit of natural inclusion was quixotic.

My exploratory interviews strongly suggested that proponents of natural inclusion were going to find it extraordinarily difficult, if not impossible, to convince busy and already overworked faculty members that they should venture beyond their disciplinary field and devote sufficient time and energy to gain requisite experience in the appropriate religious and spiritual traditions. It also became apparent that to marshal the requisite expertise and authority both to choose which issues naturally to include and to prepare faculty to include those issues in an intellectually responsible fashion was a project for a team that included committed disciplinary practitioners and scholars of religion who were willing to work with these practitioners.

On campuses where there is sufficient interest and commitment to assemble such teams, natural inclusion may work. I am far less sanguine about the likelihood of success if individual scholars attempt even circumspect natural inclusion on their own. Finally, the controversy surrounding the *Academic Bill of Rights*, which we discuss in the chapter "Academic Freedom," suggests how easily the pedagogical ideal of natural inclusion might be turned into a club to force faculty to "teach the controversy" on such issues as "intelligent design," which is controversial only because religious and political groups have made it so.

There can be good pedagogical reasons for introducing explicitly religious or spiritual claims in classroom lectures or discussions. We may wish to disclose where we are "coming from" and encourage similar self-awareness in our students. We may want our students to be aware of those issues in our discipline that are religiously contested and what the religious alternatives may be. In either case we are likely to be at least bending our discipline's standards and practices regarding religion. We may also be pushing against our students' sensibilities. We should, therefore, be able to explain and justify why we're introducing religious perspectives and why we have introduced them in the way that we have. As responsible professionals, we have additional obligations. If we decide to bring religious or spiritual convictions into the classroom, we need to be adept at handling heated exchanges that may be construed as disrespectful by one group of students or another. And if we want our students to recognize the religiously contested issues and their religious alternatives, we owe it to our students (and our colleagues in our own discipline and in Religious Studies) to handle the religious alternatives with competence and fairness. In the classroom, we are the authority. If we are going to explain religious alternatives, we need to make sure that our authority is warranted. With religion in the classroom, it is better not to venture than to venture badly.

# Conclusion

In contemporary American higher education, most of us who are scholars entered into our disciplines through a lengthy process of *professional* formation that often began in undergraduate school with our choice of majors and continued through extensive graduate training and years of probationary status as instructors and assistant professors. In the course of this long socialization, we acquired and internalized our discipline's ways of seeing, speaking, and thinking about its subject matter. We live out and reinforce this formation every day as we teach, research, and profess our discipline.

In contemporary America, those of us who belong to a religious community usually entered into it as children and went through a process of *lay* socialization that included long opportunities to observe appropriate adult behavior, periods of formal and informal instruction, and graduated opportunities to participate in the community's rituals and practices. In the course of this long socialization, we internalized our religious community's way of seeing, speaking, and thinking about self and world. And even those among us who have left the religious community in which we were formed may find, upon reflection, that the beliefs and practices of the community continue to exert an influence at a deep level.

To be a full *professional* member of a disciplinary community demands strong self-identification with the discipline; an intuitive grasp of the goods, practices, standards, and interpretive schemes of the discipline; and heavy emotional and intellectual investment. To be a full *lay* member of a religious community may demand similar engagement. In America, however, the professional formation of a disciplinary scholar is almost always more extensive and demanding than the lay formation of a modern American believer. In fact, the formation of today's academic is hardly less rigorous than the traditional formation of a nun or monk, requiring as it does extensive training, a long probationary period, years of submission to the judgment of superiors and (willing or not) to the continued judgment of peers, and, arguably, considerable intellectual asceticism and self-denial.

Finally, we belong to a third community that often bends our professional formation. Our employing college or university has its own way of shaping its members—and being shaped in turn. This is as true of "secular" colleges and universities as it is of "church-related" institutions.

Disciplinary and religious communities have powerful, coherent, and perhaps incompatible ways of construing self and world. Each selects what it sees and what it does not see; what it values and what it does not value; what it considers important, objective, and real and what it considers unimportant, subjective, and perhaps even imaginary. Each can exhibit tendencies toward totalizing explanations, overlooking, ignoring, or dismissing that which does not comport with its own views.

Today's disciplinary communities not only arose out of, but also sometimes as alternatives or even rivals to, religious communities of the nineteenth and early twentieth centuries. We faculty who are disciplinary professionals and members of a religious community often recapitulate in our own person some of the tensions inherent in this history. We must handle competing demands and operate within professional expectations that may constrain our behavior and, on occasion, pit personal and professional standards against each other.

Our accommodation takes various forms. Some of us compartmentalize our lives and move between our religious community and our disciplinary community mentally "switching gears" to suit the new context. Some others may develop "conceptual bridging strategies" that attempt to capture high-level commonalities among the different communities to which we belong. Still others live with greater ambiguity or even inconsistency in our lives. Most of us religious academics combine "gear switching" with "high-level bridging strategies," tolerate varying degrees of inconsistency, and engage in a good deal of imaginative improvisation along the way.

In seeking a viable accommodation for ourselves, our disciplines, and our colleges or universities, we faculty need to be aware of how disciplinary and religious communities have formed us profoundly, establishing in us fundamental dispositions, background beliefs, and habits of mind that often influence us without our awareness. This is especially true in matters regarding morality, human nature, and "maximally comprehensive views of reality" (i.e., metaphysics). Disciplinary and religious perspectives commonly differ in these three domains, with attendant difficulties for those of us who are both religious (or spiritual) and disciplinary professionals. The differences can also cause problems for colleagues who seek to understand us and our scholarship and teaching. The disagreements on these points are not easily settled, and, in many cases, the best we may hope for is better mutual understanding.

But even in these limited domains, we must proceed cautiously. We must always keep in mind that the collective enterprise of professional disciplinary scholarship places limits and constraints on each of us—and for good reason.

As disciplinary specialists, we are granted considerable intellectual authority in our specialty because we have undergone rigorous training and certification, and also because we continue to submit ourselves to extensive peer review, critique, and, if necessary, correction. As professionals, we are granted considerable freedom and autonomy by the larger society because we submit collectively to stringent self-regulation. The larger society will support our academic freedom so long as we hold up our end of the social bargain. Students do not enjoy this broad academic freedom because they have not yet earned it. Their limited academic freedom derives from our professional responsibility as faculty to educate them fairly and properly. So the stakes are high. If our disciplinary community does not collectively maintain what it takes to be high scholarly standards, others will impose their own standards. If our community doesn't self-regulate, others will regulate for us. If we seem not to be treating our students professionally, others will attempt to force us to do so.

These considerations bear directly on any attempt to reintroduce religious or spiritual perspectives into scholarship or teaching. If we ignore our disciplinary community's views regarding the introduction of explicitly religious perspectives into scholarship and teaching, our disciplinary peers may take action against us for violating these standards. This may strike some as an unfair limitation on individual academic freedom, but the social rationale behind professional disciplinary scholarship combined with the history of tension and conflict between religion and scholarship give our disciplinary communities ample grounds to argue otherwise. Those who want to change the standards must understand the standards' origins, rationale, and continuing importance and effectiveness.

Communities of practice have a history and are bearers of a tradition. A constituting part of that history and tradition is an ongoing argument over what the appropriate goods, standards, and practices of the community ought to be.[1] This is especially true of disciplinary communities. The goods, standards, and practices are always being put to the question, challenged, and sometimes—when the time is right and the reasons good—even changed. Now may be an apt time in our history for a circumspect reintroduction of religious perspectives into the discourse of the academy—but only if it is done right.

Mindful of such considerations, we faculty need to exercise prudence and restraint if we decide to introduce religious considerations into our scholarship, teaching, and conversation with colleagues. We must guard against repetition of the unhappy history in American higher education of religious intolerance, discrimination, and constraints on free inquiry and expression. We must respect the sensibilities of our students who rely on our professional restraint. We must remember that we are professionals rightly answerable to our larger disciplinary communities, which foster and protect a worthy and important enterprise. If we want change, we must convince colleagues by word and deed that the change truly advances the community's pursuit of knowledge and understanding in a pluralistic and religiously conflicted world.

Yet in the end, there are times when we need to be frank. There are times when we are simply obliged as disciplinary professionals to voice the explicitly religious considerations that motivate us and inform the scholarly arguments we advance. Well-considered and prudent frankness is not merely about being honest to oneself—although it is certainly that. It is not merely about letting students and colleagues know "where we are coming from"—although it can be that as well. In the exercise of careful candor, we serve in the best and deepest sense the calling that is ours as disciplinary professionals—to deepen understanding and to advance knowledge for our own time and for generations to come.

# Appendix 1

# Advice for Seminar Leaders

*Religion on Our Campuses* was originally researched and written to be the basis for faculty seminars. In order to be accessible to a larger audience, chapters were rewritten and expanded to allow the book to be helpful apart from faculty conversations. But the book or portions of it can still be used to set up faculty conversations. Appendix 2 supplies a brief stand-alone setup for "Narrative Identity."

When the conversational approach advocated in this book becomes actual conversations in a faculty seminar, it is prudent not only to describe the elements of a good conversation but also to do a bit of prescribing.[1] Conversations involve taking turns, listening carefully, giving feed back on what one hears, honoring emotions as well as ideas. When arranging for *conversations* on deep (often religious) convictions, the seminar leader will need to urge participants to take turns, listen carefully, feed back what they hear, and honor feelings and sentiments as well as ideas. When the conversation revolves around a subject as sensitive and risky as deep personal convictions, it is especially important that participants lean toward the ideal and avoid the obvious pitfalls that give rise to discord and contention rather than mutual understanding.

A few additional guidelines may help.

First, it is a good idea to pass over questions of who is right and who is wrong in favor of exploring each other's stories. Several of the chapters encourage individual narrative, and the more your group is able to share each other's stories, the more the benefit you are likely to derive from the conversation. And as you explore each other's stories, ask yourself why each of you see the world differently. What is it about your experiences, background, and differing as well as overlapping contexts that may account for your differences? Cultivate your social curiosity and let your critical judgment relax a bit!

As you converse, also listen for the feelings behind the assertions, both in your interlocutor and in yourself. Recognize that deep convictions of a religious sort tap into emotions that may make it difficult to listen to alternatives. Conversations on deep convictions can also be threatening to self-esteem and identity. Be on the lookout for reactions that make you either defensive or aggressive. Ask yourself, what is going on? Why this reaction on this topic? And remember that your interlocutors may also be experiencing strong feelings. Their identities may be at as much risk as your own.

Finally, be aware that where there are power imbalances, risk is compounded. If the group includes faculty of different ranks, the more senior faculty must be mindful of the greater risk that such conversations may pose to junior members. Differences in gender, race, ethnicity, and sexual orientation may also make some participants more wary than others. And naturally, in conversations on religion on campus, religious differences will themselves play a major role. There may be no hierarchy in a proper conversation, but that does not erase the hierarchy that participants bring to the conversation. Reticence is an understandable reaction in situations of unequal power and status. It is probably best to be explicit about this, and then to accommodate the individual decisions that interlocutors make.

The eleven chapters in this book can be used for a week-long faculty colloquium, or an intense two- or three-day retreat, or a weekly or monthly faculty seminar. Individual chapters or portions of it may also be used to set up stand-alone conversations. If time permits, each of the eleven chapters can be used to set up more than one faculty conversation, or several may be combined or omitted to fit the schedule and the focus of the seminar.

Most faculty will find that successful conversations need at least eight or nine participants and probably no more than fourteen or fifteen. These considerations probably set lower and upper limits on the size of an ideal conversation on these issues. The seminar should comprise as many disciplines as possible and (ideally) include participants from the natural sciences, social sciences, and humanities (including the arts). A productive seminar depends on securing both diverse perspectives and a safe atmosphere for all participants.

Many schedules are possible, from the leisurely to the intense. The conversations may be spread over a semester, with, say, an hour or an hour-and-a-half meeting once a week. The conversation setups could then be supplemented with other readings, the conversation on some chapters spread over more than one session, and some chapters repeated with an opportunity for individual reflection and journaling between sessions. Or the conversations might be concentrated into two or three days following a retreat format (on or off campus) where one session follows another, with breaks for food and informal exchanges. The chapters in part I (as a unit) and the stand-alone chapter "Narrative Identity" have each been used successfully for a one-day retreat.

Whatever the arrangement, the series of conversations should aim to give both adequate time for faculty to think through the questions as they apply to their own scholarly and personal lives and adequate opportunity for them to share their thoughts and stories with each other. There needs to be enough time to surface and affirm the great variety that will be found in most faculties and to take seriously the telling concerns that many have about mixing religious or spiritual claims with scholarship and teaching.

# Appendix 2

# How and Why I Became an Academic?

What story might we tell if asked how and why we chose our academic career, our disciplinary field, our research specialty, and our teaching areas?

In the modern West, most narratives on this question would include the following:

1. **Ability and incompetence** An account of the abilities that enabled us to do well (or at least well enough) to become a scholar and a teacher within our discipline and specialty, and perhaps also a recitation of some key skills or aptitudes that we lacked, thus foreclosing alternatives.
2. **Interests and aversions** An account of interests that motivated us in our choice of an academic career and in our choice of field and specialty, and of aversions that may turned us away from alternatives.
3. **Values** An account of the values that motivated us in our choice of an academic career and in our choice of field and specialty, including, perhaps, a desire through this profession to serve others or some larger cause. A sense of duty often motivates choice, whether the duty is to God or to one's "true self," or to some other obliging power. Strong values may also preclude choices that might otherwise suit us.
4. **Opportunities and obstacles** An account of key opportunities that assisted us and significant obstacles that hindered us in our professional journey. The narrative will likely include both contingent elements—"happy accidents" that favored us and obstacles we've had to overcome—and intentional interventions for and against this career choice by parents, friends, faculty, and others.

5. **Choices** An account of the key choices we made along the way that, at least in retrospect, look like turning points in our story.
6. **Mentors and role models** An account of key people who encouraged or discouraged us, who illustrated how it could be, or alternatively, should not be done.
7. **Confirmations** An account of how events confirmed or validated our choices. The narrative may include decision to admit us by *the* graduate school we wanted to attend, our college's or university's decision to hire us out of a competitive field of applicants, a key foundation grant, a prize for the best article or book, and so on. The confirmation may also be internal, a sense of "fit," perhaps even the outcome of reflection on our sense of alignment (see the next point).

Finally, there is one further consideration that can (but not necessarily) add a religious or spiritual dimension to an otherwise secular narrative:

8. **Alignment** An account that aligns our story with some larger narrative or deeper purpose. We may include among our narrative's plot elements the story of how we felt "called" to our career by something larger than ourselves. Or we might tell of how we gradually discovered our "true self" or our "deep identity" and learned to live that out through our life and work. Or we might narrate how we discovered an almost mystically fine fit between our abilities, interests, and values and the requirements of the profession we have chosen.

## Questions for Reflection and Discussion

Ask yourself: In any account I might give of how I became a scholar, disciplinary professional, and teacher, what mention would be made of the following:

1. **Abilities** Would my narrative include an account of my intellectual strengths and weaknesses, or of my abilities generally? How do I account for these abilities? What role did they play in the shaping of my career? Do they in any way spring from, or relate to, my core convictions? Do I view my abilities as gifts, either literally or figuratively? In what ways did my intellectual strengths and weaknesses particularly fit me well for some careers, but not others? Would I see my abilities as "evidence" that I may have been in some sense "called" to a career that demands such skills?

2. **Interests** Would my narrative include an account of my major interests and aversions? When and how did they arise? How did they influence my choices about career, scholarship, and teaching? Am I studying and teaching materials that in some way deeply resonate with my core convictions and beliefs? Am I studying and teaching about matters in part because I have personal as well as professional questions about them?
3. **Values** Would my narrative include an account of how my core convictions influenced my career choices? At key points did they rule out some options and encourage others? Did I choose my field, my specialty, or my research topic because it allowed me to study something that I greatly valued or that was central in my life? Did I choose my profession out of a sense of duty, and if so, duty to what or whom? Did I or do I view my profession as a means of service to others or to some larger cause?
4. **Opportunities and obstacles** What were the major opportunities that facilitated my career and the obstacles that either hindered it or steered me in a certain direction? Were these opportunities and obstacles largely fortuitous or did they result from the action of others such as mentors, senior faculty, friends and family, rivals and competitors?
5. **Choices** What were the key choices I made that shaped my career? What was the basis for these choices? In what ways were these choices free and in what ways were they constrained? In what ways did my core convictions inform or determine these choices?
6. **Confirmations** How have key choices or turning points in my career been confirmed or validated by events? Are there examples of disconfirmation as well? How have I understood and dealt with confirmation and disconfirmation?
7. **Alignment** Do I in any way feel "called" to my profession? If so, did the call (or calls) arise from within or from without? Do I feel particularly "fitted" to my career by my abilities, interests, and opportunities? Am I through my career attempting to serve some larger cause, whether religious, spiritual, philosophical, political, or social?
8. **Ambiguities and complexities** Does my narrative do justice to the ambiguities, uncertainties, doubts, and dissatisfactions that I have experienced in my career?
9. **Communities** What role has the various communities of which I'm a member—including my disciplinary community and, if I have one, my religious community—played in my narrative?

What is my *primary* community in this regard? Why is it my primary community? How have the various communities either reinforced or conflicted with each other?

10. **Fooling ourselves?** Some suggest that retrospective narratives often invest happenstance with unwarranted significance and allow us to invent meaning by tendentious selection and even distortion. How valid or plausible do I find this critique when applied to my own narrative and to that of others?

# Notes

## Introduction

### Notes to Pages 2–6

1. Catholic colleges and universities have their own challenges, commonly involving difficult relations with the Catholic hierarchy. See David J. O'Brien, *From the Heart of the American Church: Catholic Higher Education and American Culture* (Maryknoll, NY: Orbis, 1994).
2. The example is drawn from Richard Rorty who sees the mention of religious conviction in political discussion as a conversation-stopper. The same considerations apply to discussions generally within the academy. See Richard Rorty, "Religion as Conversation-Stopper (1994)," in *Philosophy and Social Hope* (London: Penguin Books, 1999), 168–74.
3. For a highly insightful discussion of how to handle a difficult conversation, see Douglas Stone, Bruce Patton, and Sheila Heen, *Difficult Conversations: How to Discuss What Matters Most* (New York: Viking, 1999).
4. I say in the sense analogous to that intended by Oakeshott because Oakeshott was advocating and describing a "conversation" among modes of understanding, not among individuals. For an elaboration and interpretation of his philosophical teachings and his view of conversation among modes of understanding, see Paul Franco's *Michael Oakeshott: An Introduction* (New Haven, CT: Yale University Press, 2004) and Terry Nardin's *The Philosophy of Michael Oakeshott* (University Park, PA: The Pennsylvania State University Press, 2001).
5. Michael Oakeshott, "The Voice of Poetry in the Conversation of Mankind (1962)," in *Rationalism in Politics and Other Essays* (Indianapolis: Liberty Fund, 1991), 489–90. I am deploying here (for different purposes) a way of framing the issue for which I am indebted to Martin E. Marty, *The One and the Many: America's Struggle for the Common Good* (Cambridge, MA: Harvard University Press, 1997).
6. Marty, *The One and the Many*, 22–23, 154–60.
7. Oakeshott, "The Voice of Poetry," 490.
8. Ibid.
9. The literature on this is vast. See, among others, Robert N. Bellah, Richard Madsen, William M. Sullivan, Ann Swidler, and Steven M. Tipton, *The Good Society* (New York: Vintage, 1991); Robert N. Bellah, Richard Madsen, William M. Sullivan, Ann Swidler, and Steven M. Tipton, *Habits of the Heart: Individualism and Commitment in American Life* (Berkeley: University of California Press, 1985); Richard J. Bernstein, *Beyond Objectivism and Relativism: Science,*

*Hermeneutics, and Praxis* (Philadelphia: University of Pennsylvania Press, 1983); Alasdair MacIntyre, *After Virtue: A Study in Moral Theory*, 2nd ed. (South Bend, IN: University of Notre Dame Press, 1984); Alasdair MacIntyre, *Three Rival Versions of Moral Enquiry: Encyclopaedia, Genealogy, and Tradition* (South Bend, IN: University of Notre Dame Press, 1990); Alasdair MacIntyre, *Whose Justice? Which Rationality*? (Notre Dame, IN: University of Notre Dame Press, 1988); Jeffrey Stout, *Democracy and Tradition* (Princeton, NJ: Princeton University Press, 2004); Jeffrey Stout, *Ethics After Babel: The Languages of Morals and Their Discontents* (Boston: Beacon Press, 1988); Charles Taylor, *Sources of the Self: The Making of the Modern Identity* (Cambridge, MA: Harvard University Press, 1989); and Michael Walzer, *Thick and Thin: Moral Argument at Home and Abroad* (Notre Dame, IN: University of Notre Dame Press, 1994).

10. Christian Smith and Melinda Lundquist Denton, *Soul Searching: The Religious and Spiritual Lives of American Teenagers* (New York: Oxford University Press, 2005), 143–44.
11. For a brief overview and a guide to the literature, see *Communitarianism* [Web] (Wikipedia, July 16, 2005 [cited July 22, 2005]); available from http://en.wikipedia.org/wiki/Communitarianism.
12. For a most recent, comprehensive survey, see Barry A. Kosmin, Egon Mayer, and Ariela Keysar, *American Religious Identification Survey 2001* (New York: The Graduate Center of the City University of New York, 2001).
13. I am not aware of a study on the religious self-identification of faculty members. UCLA's Higher Education Research Institute is planning as part of its *Spirituality in Higher Education* project to remedy this lack. For results on its student survey, see Alexander Astin and Helen Astin, "The Spiritual Life of College Students: A National Study of College Students' Search for Meaning and Purpose," in *Spirituality in Higher Education* (Los Angeles: UCLA Higher Education Research Institute, 2005); Alexander Astin and Helen Astin, "The Spiritual Life of College Students: A National Study of College Students' Search for Meaning and Purpose. Executive Summary" (Los Angeles: UCLA Higher Education Research Institute, 2005). These two reports and supporting documents are available on the project's web site: http://www.spirituality.ucla.edu/reports/index.html.
14. This is Richard Bernstein's label for a concept that goes back to the American pragmatist Charles Sanders Peirce. Richard J. Bernstein, *The New Constellation: The Ethical—Political Horizons of Modernity Postmodernity* (Cambridge, MA: The MIT Press, 1992), 48, 328, 36.
15. Thomas L. Haskell, *The Emergence of Professional Social Science: The American Social Science Association and the Nineteenth-Century Crisis of Authority* (Urbana, IL: The University of Illinois Press, 1977; reprint, Johns Hopkins University Press, 2000), 18–19.

## 1 Cautionary Tales

1. The following account is drawn from Jon H. Roberts and James Turner's *The Sacred and the Secular University* (Princeton, NJ: Princeton University Press, 2001).

2. Ibid., 23.
3. Ibid., 29.
4. Ibid., 47.
5. Julie A. Reuben, *The Making of the Modern University: Intellectual Transformation and the Marginalization of Morality* (Chicago: The University of Chicago Press, 1996), 176.
6. See Bruce A. Kimball, *Orators and Philosophers: A History of the Idea of Liberal Education* (New York: Teachers College Press, 1986).
7. James Turner, *The Liberal Education of Charles Eliot Norton* (Baltimore: Johns Hopkins University Press, 1999).
8. See Roberts and Turner, *Sacred and Secular University*. For a survey of the issues and an introduction to the vast literature spawn by the "culture wars" and "the battle over the canon," see http://chronicle.com/indepth/culture/canon.htm. For useful background, see W. B. Carnochan, *The Battleground of the Curriculum: Liberal Education and American Experience* (Stanford, CA: Stanford University Press, 1993) and Kimball, *Orators and Philosophers*.
9. These tales can often also be classified under the heading "narratives of discrimination."
10. Reuben, *Making of the Modern University*, 206.
11. You'll find this argument repeated even today. See, e.g., Jeffrey Stout, *Ethics After Babel: The Languages of Morals and Their Discontents* (Boston: Beacon Press, 1988), 222.
12. Mark R. Schwehn, among others, challenges the equation of religion and violence. See his *Exiles from Eden: Religion and the Academic Vocation in America* (New York: Oxford University Press, 1993), 30.
13. Reuben, *Making of the Modern University*, 187–88, quote on p. 188.
14. I employ George Marsden's definition of established liberal Christianity within higher education: "By 'liberal Protestant,' " Marsden explains, "I mean a culture that took for granted Protestantism as one significant part of its heritage, but was 'liberal' in that it emphasized the unifying moral dimensions of its spiritual heritage, rather than the particulars of traditional Protestant doctrine. Today's pace-setting American universities were virtually all constructed in the late nineteenth and early twentieth century by liberal Protestants." George M. Marsden, *The Outrageous Idea of Christian Scholarship* (New York: Oxford University Press, 1997), 14.
15. Of course, they also discriminated against women, who were for many years excluded from undergraduate admissions to the most exclusive elite universities, and against blacks, who were discriminated against at most American colleges and universities.
16. See George M. Marsden, *The Soul of the American University: From Protestant Establishment to Established Nonbelief* (New York: Oxford University Press, 1994), 357–66, who cites Marcia Graham Synnott, *The Half-Opened Door: Discrimination and Admissions at Harvard, Yale, and Princeton, 1900–1970* (Westport, CT: Greenwood Press, 1979).
17. Marsden, *Soul of the American University*, 363, citing Synnott, *The Half-Opened Door*, 15–16.

18. Dan A. Orien, *Joining the Club: A History of Jews and Yale* (New Haven, CT: Yale University Press, 1985), 175–76.
19. Marsden, *Soul of the American University*, 357.
20. See ibid., 357ff. and David A. Hollinger, *Science, Jews, and Secular Culture: Studies in Mid-Twentieth-Century American Intellectual History* (Princeton, NJ: Princeton University Press, 1996), 18–23.
21. See Hollinger, *Science, Jews, and Secular Culture*, 7–8 who cites Orien, *Joining the Club*.
22. Hollinger, *Science, Jews, and Secular Culture*, 8.
23. See Marsden, *Soul of the American University*, 365 and Orien, *Joining the Club*, 120–21. For background reading, see Reuben, *Making of the Modern University*; Roberts and Turner, *Sacred and Secular University*; James Turner, "Secularization and Sacralization: Speculations on Some Religious Origins of the Secular Humanities Curriculum, 1850–1900," in *The Secularization of the Academy*, ed. George M. Marsden and Bradley J. Longfield (New York: Oxford University Press, 1992), 74–106.
24. Marsden, *Soul of the American University*, 400–04. See also Michael J. Buckley, S.J., *The Catholic University as Promise and Project* (Washington, DC: Georgetown University Press, 1998); Philip Gleason, "American Catholic Higher Education, 1940–1990," in *The Secularization of the Academy*, ed. George M. Marsden and Bradley J. Longfield (Oxford: Oxford University Press, 1992), 234–58; and David J. O'Brien, *From the Heart of the American Church: Catholic Higher Education and American Culture* (Maryknoll, NY: Orbis, 1994).
25. Orien, *Joining the Club*, 120–21.

## 2 Encounters

1. See the discussion in "Cautionary Tales." Dan Orien has, e.g., explored how Jews were discriminated against at Yale University well into the mid-twentieth century. See Dan A. Orien, *Joining the Club: A History of Jews and Yale* (New Haven, CT: Yale University Press, 1985).
2. John Witte, Jr. and Richard C. Martin, eds., *Sharing the Book: Religious Perspectives on the Rights and Wrongs of Proselytism* (Maryknoll, NY: Orbis, 1999).
3. Martin E. Marty, "Introduction: Proselytizers and Proselytizees on the Sharp Arête of Modernity," in *Sharing the Book: Religious Perspectives on the Rights and Wrongs of Proselytism*, ed. John Witte, Jr. and Richard C. Martin (Maryknoll, NY: Orbis, 1999), 1–14, quote on p. 1.
4. Christian Smith, *American Evangelicalism: Embattled and Thriving* (Chicago: The University of Chicago Press, 1998), 181.
5. Marty, "Proselytizers and Proselytizees," 2.
6. How truly free such choices actually are is a matter of some dispute. See parts I and II where we explore how choice is constrained by the communities to which we belong.

7. See, e.g., Roger Kimball, *Tenured Radicals: How Politics Has Corrupted Our Higher Education* (New York: Harper & Row, 1990), and Charles J. Sykes, *Profscam: Professors and the Demise of Higher Education* (New York: Regnery Gateway, 1988).
8. According to the ideals of the academy, by way of contrast, the educator employs doubt as a means to make way for a more sophisticated ways of handling ongoing doubt.
9. Marty, "Proselytizers and Proselytizees," 3.
10. J. Budziszewski, *How to Stay Christian in College* (Colorado Springs, CO: Th1nk Books, 2004); Tony Campolo and William Willimon, *The Survival Guide for Christians on Campus: How to Be Students and Disciples at the Same Time* (West Monroe, LA: Howard Publishing Company, 2002).
11. These examples are section headings from Budziszewski, *How to Stay Christian.*
12. Although I would not agree with Budziszewski on many points, his book, from which I've drawn these section headings, falls into the thoughtful and responsible category.
13. Students for Academic Freedom, *Academic Bill of Rights* (2004 [cited June 28, 2005]); available from http://www.studentsforacademicfreedom.org/abor.html.
14. The quoted phrase is drawn from the American Association of University Professors (AAUP) 1967 *Joint Statement on Rights and Freedoms of Students.* AAUP, *Joint Statement on Rights and Freedoms of Students* (1967 [cited June 27, 2005]); available from http://www.aaup.org/statements/Redbook/Studentrights.htm.
15. A similar conclusion was reached by the authors of the 1915 AAUP *Declaration of Principles.* See AAUP, "The 1915 Declaration of Principles," in *Academic Freedom and Tenure: A Handbook of the American Association of University Professors,* ed. Louis Joughin (Madison: University of Wisconsin Press, 1969), 170, and Neil Hamilton, *Zealotry and Academic Freedom: A Legal and Historical Perspective* (New Brunswick, NJ: Transaction Publishers, 1995), 366.

## 3 Religious Formation

1. Many of the enduring characteristics of the modern scholarly profession were described by the great analyst of modernity, Max Weber, in his "Wissenschaft als Beruf" (somewhat misleadingly translated as "Science as a Vocation"). See Max Weber, "Science as a Vocation," in *From Max Weber: Essays in Sociology,* ed. H. H. Gerth and C. Wright Mills (New York: Oxford University Press, 1977), 129–56. On this point see the excellent discussion in Mark R. Schwehn's *Exiles from Eden: Religion and the Academic Vocation in America* (New York: Oxford University Press, 1993).
2. William E. Paden, *Interpreting the Sacred: Ways of Viewing Religion* (Boston: Beacon Press, 1992), 6.
3. For my definition, I rely principally on George A. Lindbeck, *The Nature of Doctrine: Religion and Theology in a Postliberal Age* (Philadelphia: The Westminster Press, 1984); Alasdair MacIntyre, *After Virtue: A Study in Moral Theory,* 2nd ed. (South Bend, IN: University of Notre Dame Press, 1984);

Paden, *Interpreting the Sacred*, Ronald F. Thiemann, *Constructing a Public Theology* (Louisville, KY: Westminster/John Knox Press, 1991); Ronald F. Thiemann, *Religion in Public Life: A Dilemma for Democracy, A Twentieth Century Fund Book* (Washington, DC: Georgetown University Press, 1996); and Ronald F. Thiemann, *Revelation and Theology: The Gospel as Narrated Promise* (Notre Dame, IN: University of Notre Dame Press, 1985).

4. William E. Paden, *Religious Worlds: The Comparative Study of Religion* (Boston: Beacon Press, 1988, 1994), 11.
5. Ibid., 10.
6. Lindbeck, *The Nature of Doctrine*, 33.
7. Ibid.
8. Ibid., 34.
9. Paden, *Interpreting the Sacred*, 9.
10. Lindbeck, *The Nature of Doctrine*, 35.
11. Ibid., 32.
12. See also Paden's comment about the breadth of religion: "Religious systems are designed to shape the overall way one perceives and construes existence. They do not merely define some limited realm of behavior within the world." Paden, *Religious Worlds*, 11.
13. Thiemann, *Religion in Public Life*, 132.
14. Lindbeck, *The Nature of Doctrine*, 32.
15. Thiemann, *Religion in Public Life*, 132.
16. For this brief description of spirituality and spiritual practices, I have followed Robert Wuthnow, "Spirituality and Spiritual Practice," in *The Blackwell Companion to Sociology of Religion*, ed. Richard K. Fenn (Malden, MA: Blackwell Publishers, 2001), 306–20, who, in turn, has adapted Alasdair MacIntyre's understanding of social practices (MacIntyre, *After Virtue*, 187). For some alternative views, see Robert N. Bellah, Richard Madsen, William M. Sullivan, Ann Swidler, and Steven M. Tipton, *Habits of the Heart: Individualism and Commitment in American Life* (Berkeley: University of California Press, 1985); Robert C. Fuller, *Spiritual But Not Religious: Understanding Unchurched America* (New York: Oxford University Press, 2001); and Christian Smith and Melinda Lundquist Denton, *Soul Searching: The Religious and Spiritual Lives of American Teenagers* (New York: Oxford University Press, 2005).
17. On this choice, see Tom Beaudoin, *Virtual Faith: The Irreverent Spiritual Quest of Generation X* (San Francisco: Jossey-Bass, 1998); Fuller, *Spiritual But Not Religious*; George Gallup, Jr. and Timothy Jones, *The Next American Spirituality: Finding God in the Twenty-First Century* (Colorado Springs, CO: Victor, 2000); Wade Clark Roof, *A Generation of Seekers: The Spiritual Journeys of the Baby Boom Generation* (San Francisco: HarperSanFrancisco, 1993); Wade Clark Roof, *Spiritual Marketplace: Baby Boomers and the Remaking of American Religion* (Princeton, NJ: Princeton University Press, 1999); Robert Wuthnow, *After Heaven: Spirituality in America since the 1950s* (Berkeley, CA: University of California Press, 1998). For some probing questions regarding at least the teenage generation, see Smith and Denton, *Soul Searching*. For some questions about the individualism that may underlie this choice, see especially Robert N.

Bellah, Richard Madsen, William M. Sullivan, Ann Swidler, and Steven M. Tipton, *The Good Society* (New York: Vintage, 1991) and Bellah et al., *Habits of the Heart.*

18. Conrad Cherry, Betty A. DeBerg, and Amanda Porterfield, *Religion on Campus: What Religion Really Means to Today's Undergraduates* (Chapel Hill: The University of North Carolina Press, 2001), 276–77.
19. Beaudoin, *Virtual Faith*, 45, 178. Cited in Cherry et al., *Religion on Campus*, 276.
20. Beaudoin, *Virtual Faith*; Roof, *Generation of Seekers*; Roof, *Spiritual Marketplace.*
21. Wuthnow, "Spirituality and Spiritual Practice," 319. Wuthnow, *After Heaven.*
22. Robert Wuthnow, *Sharing the Journey: Support Groups and America's New Quest for Community* (New York: Free Press, 1994).
23. This is often explained by sociologists and economists (rational choice theorists) in terms of "self-fashioning" and identity formation through a process of selective consumption. For a review of the literature, see David Lyon, *Jesus in Disneyland: Religion in Postmodern Times* (Oxford: Polity Press, 2000).
24. Paul J. Griffiths, "Reading as a Spiritual Discipline," in *The Scope of Our Art: The Vocation of the Theological Teacher*, ed. L. Gregory Jones and Stephanie Paulsell (Grand Rapids, MI: Eerdmans, 2002); Paul J. Griffiths, *Religious Reading: The Place of Reading in the Practice of Religion* (Oxford: Oxford University Press, 1999); and Stephanie Paulsell, "Writing as a Spiritual Discipline," in *The Scope of Our Art: The Vocation of the Theological Teacher*, ed. L. Gregory Jones and Stephanie Paulsell (Grand Rapids, MI: Eerdmans, 2002), 17–31.
25. The notion of "orientation" as used here entails mental and emotional "set or attitude," a given range of imagination and sympathy, an orientation captured, perhaps, by the German *Einstellung.*
26. In the following I draw on, among others, Peter L. Berger, ed., *The Desecularization of the World: Resurgent Religion and World Politics* (Grand Rapids, MI: Eerdmans, 1999); David Fontana, *Psychology, Religion, and Spirituality* (Malden, MA: Blackwell Publishers, 2003); Lyon, *Jesus in Disneyland*; and Wuthnow, "Spirituality and Spiritual Practice."
27. Peter L. Berger, *A Far Glory: The Quest for Faith in an Age of Credulity* (New York: Free Press, 1992), 66.
28. Berger calls this option "cognitive bargaining." Ibid., 41.
29. See the brief discussion of *bricolage* in the chapter "Narrative Identity."
30. Stephen Jay Gould, *Rocks of Ages: Science and Religion in the Fullness of Life* (New York: Ballantine Books, 1999).
31. Ian Barbour, *Ethics in an Age of Technology: The Gifford Lectures, Volume Two* (San Francisco: HarperSanFrancisco, 1993); Ian Barbour, *Myths, Models, and Paradigms: A Comparative Study in Science and Religion* (New York: Harper & Row, 1974); Ian Barbour, *Religion and Science: Historical and Contemporary Issues* (San Francisco: HarperSanFrancisco, 1997); John Polkinghorne, *Belief in God in an Age of Science* (New Haven, CT: Yale University Press, 1998); John Polkinghorne, *Beyond Science* (Cambridge: Cambridge University Press, 1996); and John Polkinghorne, *The Faith of a Physicist, Theology and the Sciences* (Minneapolis, MN: Fortress Press, 1994).
32. See the previous section for a brief discussion of spiritual practices.

# 4 DISCIPLINARY FORMATION

1. Thomas L. Haskell, *The Emergence of Professional Social Science: The American Social Science Association and the Nineteenth-Century Crisis of Authority* (Urbana, IL: The University of Illinois Press, 1977; reprint, Johns Hopkins University Press, 2000), 18–19.
2. For simplicity of exposition, I shall continue to refer to disciplinary communities, but the reader needs to recognize that most scholars identify themselves with a subfield within a discipline. These subdisciplinary communities often blur into each other, with individual scholars straddling more than one subdisciplinary field even in their research and almost always in their teaching.
3. I am adapting here the ethicist Alasdair MacIntyre's definition of *practice* and applying it to an academic discipline. MacIntyre's famous one-sentence definition: "By a 'practice' I am going to mean any coherent and complex form of socially established cooperative human activity through which goods internal to that form of activity are realized in the course of trying to achieve those standards of excellence which are appropriate to, and partially definitive of, that form of activity, with the result that human powers to achieve excellence, and human conception of the ends and goods involved, are systematically extended." Alasdair MacIntyre, *After Virtue: A Study in Moral Theory*, 2nd ed. (South Bend, IN: University of Notre Dame Press, 1984), 187.
4. The story of this revolutionary reorientation has been well told by several historians, especially George Marsden, Julie Reuben, Thomas Haskell, Jon Roberts, and James Turner. Haskell, *Emergence of Professional Social Science*; Thomas L. Haskell, "Justifying the Rights of Academic Freedom in the Era of Power/Knowledge," in *The Future of Academic Freedom*, ed. Louis Menand (Chicago: The University of Chicago Press, 1996), 43–90; George M. Marsden, "The Collapse of American Evangelical Academia," in *Faith and Rationality: Reason and Belief in God*, ed. Alvin Plantinga and Nicholas Wolterstorff (Notre Dame, IN: University of Notre Dame Press, 1983), 219–64; George M. Marsden, *The Soul of the American University: From Protestant Establishment to Established Nonbelief* (New York: Oxford University Press, 1994); Julie A. Reuben, *The Making of the Modern University: Intellectual Transformation and the Marginalization of Morality* (Chicago: The University of Chicago Press, 1996); and Jon H. Roberts and James Turner, *The Sacred and the Secular University* (Princeton, NJ: Princeton University Press, 2001). See also Frederick Rudolph, *The American College and University: A History*, reissue ed. (Athens, GA: University of Georgia Press, 1985/1991); John R. Thelin, *History of American Higher Education* (Baltimore: Johns Hopkins University Press, 2004); and Laurence R. Veysey, *The Emergence of the American University* (Chicago: The University of Chicago Press, 1965). I draw heavily from their accounts. I urge the interested reader to consult their works for a much fuller and nuanced story. Here I shall only paint with a broad brush and primarily with the aim of setting the context for material that comes later in this book.
5. For example, the shift from theology to moral philosophy as the primary bearer of the Christian character of collegiate intellectual life reflects the colleges' (market-driven)

need to serve more than one church and thus a need to stress commonalities rather than theological differences. See Marsden, *Soul of the American University*, 99.

6. Francis Oakley, *Community of Learning: The American College and the Liberal Arts Tradition* (New York: Oxford University Press, 1992), 26. Citing Rudolph, *American College and University*, 47.
7. For a more detailed and nuanced description of the assumptions that dominated antebellum higher education, see especially Marsden, "The Collapse of American Evangelical Academia." Marsden, *Soul of the American University* and Reuben, *Making of the Modern University*.
8. The term is Theodore Dwight Bozeman's, cited in Roberts and Turner, *Sacred and Secular University*, 23.
9. Marsden, "The Collapse of American Evangelical Academia," 230–31.
10. As the Harvard moral philosopher Francis Bowen put it, moral philosophy was "a general science of Human Nature, of which the special sciences of Ethics, Psychology, Aesthetics, Politics, and Political Economy are so many departments." Cited in Reuben, *Making of the Modern University*, 20.
11. On this, see among others, Marsden, "The Collapse of American Evangelical Academia," 224–28 and Reuben, *Making of the Modern University*, 19–20, 36–39. For a masterful interpretation of Reid, see Nicholas Wolterstorff, *Thomas Reid and the Story of Epistemology*, ed. Robert B. Pippin, *Modern European Philosophy* (New York: Cambridge University Press, 2004).
12. Marsden, "The Collapse of American Evangelical Academia," 231.
13. On these main points there was little disagreement among the Christian denominations of the time, and even the Deists shared many of the same assumptions. Ibid., 230.
14. Reuben, *Making of the Modern University*, 21.
15. Cited in ibid. and Marsden, "The Collapse of American Evangelical Academia," 231.
16. This conviction regarding the essential harmony between Scripture and nature underwrote two of the most read books of their day, which were also staples in the antebellum curriculum: William Paley's *Natural Theology* (1802) and Bishop Joseph Butler's even more popular *The Analogy of Religion, Natural and Revealed* (1736). Paley offered the famous analogy of finding a watch and concluding that the watch must have a designer. Butler took the battle to the Deists who criticized the specifics of revealed Christianity. Butler countered that all the objections that might be lodged against revealed Christianity would apply with equal force to what he claimed all (i.e., Christians and Deists) would agree on regarding nature itself. Marsden, "The Collapse of American Evangelical Academia," 228–30.
17. Marsden, "The Collapse of American Evangelical Academia," 233.
18. Roberts and Turner, *Sacred and Secular University*, 29.
19. The 14 charter members were Harvard University, Johns Hopkins University, Columbia University, University of Chicago, University of California, Clark University, Cornell University, Catholic University, University of Michigan, Leland Stanford, Jr., University, University of Wisconsin, University of Pennsylvania, Princeton University, and Yale University. Thelin, *History of American Higher Education*, 110.

20. In the 10-year period between the founding of the AAU and Slosson's list, Clark University and Catholic University had fallen variously into difficulties and their places were taken by two state universities, the University of Illinois and the University of Minnesota. Ibid., 111.
21. Roberts and Turner, *Sacred and Secular University*, 35.
22. Ibid., 36.
23. Ibid., 29.
24. Simon Blackburn, ed., *The Oxford Dictionary of Philosophy* (New York: Oxford University Press, 1994), 255.
25. For a survey from one side of the "science studies" divide, see James Robert Brown, *Who Rules in Science: An Opinionated Guide to the Wars* (Cambridge, MA: Harvard University Press, 2001).
26. Paul J. Griffiths, "Reading as a Spiritual Discipline," in *The Scope of Our Art: The Vocation of the Theological*, ed. L. Gregory Jones and Stephanie Paulsell (Grand Rapids, MI: Eerdmans, 2002), 32–47, and Paul J. Griffiths, *Religious Reading: The Place of Reading in the Practice of Religion* (Oxford: Oxford University Press, 1999).
27. I owe this important qualification to Stephanie Paulsell. See Stephanie Paulsell, "Writing as a Spiritual Discipline," in *The Scope of Our Art: The Vocation of the Theological Teacher*, ed. L. Gregory Jones and Stephanie Paulsell (Grand Rapids, MI: Eerdmans, 2002), 17–31.
28. The term "master" can understandably raise hackles, and many faculty would prefer a less hierarchical relationship between an academic mentor and a PhD student. But for good or for ill, the hierarchical model suggested by the terms "master" and "apprentice" does still characterize the social relationship found in many of the country's leading doctoral programs.
29. Jean Lave and Etienne Wenger, *Situated Learning: Legitimate Peripheral Participation* (Cambridge, UK: Cambridge University Press, 1991), 33.
30. Ibid., 111.
31. The following is especially indebted to the learning stages suggested by Hubert L. Dreyfus, *On the Internet* (New York: Routledge, 2001), 33–49.
32. Jerome Bruner, *The Culture of Education* (Cambridge, MA: Harvard University Press, 1996).
33. Michael Silberstein and John McGeever, "The Search for Ontological Emergence," *The Philosophical Quarterly* 49: 195, 1999.
34. See Griffiths, "Reading as a Spiritual Discipline"; Griffiths, *Religious Reading*; and Paulsell, "Writing as a Spiritual Discipline."
35. For a discussion of "background" or "control beliefs," see the chapter "Inclinations."

## 5 Institutional Settings

1. I deal largely with colleges and universities in this chapter, but the reader needs to keep in mind that other institutional embodiments of disciplinary communities—e.g., institutes, foundations, and public and private centers of research and practice— may differ in significant ways from colleges and universities, not only

offering an informative contrast with, but also exerting a formative influence on, colleges and universities.

2. I draw the following with some modification from Alasdair MacIntyre's *After Virtue: A Study in Moral Theory*, 2nd ed. (South Bend, IN: University of Notre Dame Press, 1984).
3. Ibid., 194.
4. See the chapter "Academic Freedom."
5. Scholars recognize the problems with the term "mainline." They tend to use the definition provided by William R. Hutchison. See William R. Hutchison, ed., *Between the Times: The Travail of the Protestant Establishment in America, 1900–1960* (New York: Cambridge University Press, 1989), x.
6. Recent literature on church-related higher education is vast. Here's a sampling whose argument and associated bibliography can introduce you to the literature: For the history of church-relatedness of much of American higher education, see especially George M. Marsden, *The Soul of the American University: From Protestant Establishment to Established Nonbelief* (New York: Oxford University Press, 1994); Julie A. Reuben, *The Making of the Modern University: Intellectual Transformation and the Marginalization of Morality* (Chicago: The University of Chicago Press, 1996); Jon H. Roberts and James Turner, *The Sacred and the Secular University* (Princeton, NJ: Princeton University Press, 2001); and Douglas Sloan, *Faith and Knowledge: Mainline Protestantism and American Higher Education* (Louisville, KY: Westminster/John Knox Press, 1994). For an evaluation of current conditions, see the (sometimes conflicting) assessments in Robert Benne, *Quality with Soul: How Six Premier Colleges and Universities Keep Faith with Their Religious Traditions* (Grand Rapids, MI: Eerdmans, 2001); James Tunstead Burtchaell, *The Dying of the Light: The Disengagement of Colleges and Universities from the Christian Churches* (Grand Rapids, MI: Eerdmans, 1998); Conrad Cherry, Betty A. DeBerg, and Amanda Porterfield, *Religion on Campus: What Religion Really Means to Today's Undergraduates* (Chapel Hill: The University of North Carolina Press, 2001); Merrimon Cuninggim, *Uneasy Partners: The College and the Church* (Nashville, TN: Abingdon Press, 1994); and Paul J. Dovre, ed., *The Future of Religious Colleges: The Proceedings of the Harvard Conference on the Future of Religious Colleges, October 6–7, 2000* (Grand Rapids, MI: Eerdmans, 2002).
7. Cuninggim, *Uneasy Partners*, 43–45.
8. See, e.g., ibid.; Mark U. Edwards, Jr., " 'Dying of the Light' at Christian Colleges—or a Different Refraction?," *Christian Century*, April 21–28, 1999, 459–63; and David J. O'Brien, *From the Heart of the American Church: Catholic Higher Education and American Culture* (Maryknoll, NY: Orbis, 1994).

## 6 NARRATIVE IDENTITY

1. See, e.g., John R. Searle, *Mind, Language, and Society: Philosophy in the Real World* (New York: Basic Books, 1998), and John R. Searle, *Mind: A Brief Introduction* (New York: Oxford University Press, 2004).

2. For more on this, see the section on proselytizing in the chapter "Reticence."
3. Rick Warren, *The Purpose Driven Life: What on Earth Am I Here For?* (Grand Rapids, MI: Zondervan, 2002).
4. This section draws from several helpful essays, especially the works of Nancy Ammerman, Elinor Ochs, and Lisa Capps. Nancy T. Ammerman, "Religious Identities and Religious Institutions," in *Handbook for the Sociology of Religion*, ed. Michele Dillon (Cambridge, UK: Cambridge University Press, 2003); Jerome Bruner, *Acts of Meaning* (Cambridge, MA: Harvard University Press, 1990); Jerome Bruner, *The Culture of Education* (Cambridge, MA: Harvard University Press, 1996); Jerome Bruner, *Making Stories: Law, Literature, Life* (New York: Farrar, Straus and Giroux, 2002); Erving Goffman, *The Presentation of Self in Everyday Life* (Garden City, NY: Doubleday, 1958); Martin E. Marty, *The One and the Many: America's Struggle for the Common Good* (Cambridge, MA: Harvard University Press, 1997); Terry Nardin, *The Philosophy of Michael Oakeshott* (University Park, PA: The Pennsylvania State University Press, 2001); Elinor Ochs and Lisa Capps, *Living Narrative: Creating Lives in Everyday Storytelling* (Cambridge, MA: Harvard University Press, 2001); and Margaret R. Somers, "The Narrative Constitution of Identity: A Relational and Network Approach," *Theory and Society* 23 (1994), 605–49.
5. Ochs and Capps, *Living Narrative*, 2.
6. Ibid., 7.
7. Ibid., 2–3.
8. "Born-again" narratives within segments of the Christian tradition illustrate the interacting role of individual and community in shaping narrative identity. The community offers a plot outline. The individual narrator supplies the details. The community of believers help the individual fit details to the plot, and thereby become the coauthors of this key piece of religious identity.
9. Ochs and Capps, *Living Narrative*, 4.
10. See especially chapter 4 in Bruner's *Acts of Meaning*.
11. See the discussion of "unmasking" in "Cautionary Tales."
12. Bruner, *Making Stories*, 15.
13. Ochs and Capps, *Living Narrative*, 45.
14. These elements might be seen as elements of a "worldview." See, e.g., Richard J. Mouw, "Christian Scholarship: The Difference a Worldview Makes," *The Cresset*, Special Lilly issue (1997), 5–15. Alternatively, they may be categorized as "control beliefs." See Nicholas Wolterstorff, *Reason within the Bounds of Religion*, 2nd ed. (Grand Rapids, MI: Eerdmans, 1984). The point from my perspective is that context can determine applicability.
15. Borrowing from Claude Lévi-Strauss, who is far more critical of this practice than either Stout or I. See Jeffrey Stout, *Ethics After Babel: The Languages of Morals and Their Discontents* (Boston: Beacon Press, 1988), 74–77.
16. Raymond E. Brown, *The Birth of the Messiah: A Commentary on the Infancy Narratives in Matthew and Luke* (Garden City, NY: Image Books, 1977).
17. Parker J. Palmer, *Let Your Life Speak: Listening for the Voice of Vocation* (San Francisco: Jossey-Bass, 2000), 2.

18. See, e.g., the discussion in David Damrosch, *We Scholars: Changing the Culture of the University* (Cambridge, MA: Harvard University Press, 1995).
19. Mark R. Schwehn, *Exiles from Eden: Religion and the Academic Vocation in America* (New York: Oxford University Press, 1993).
20. Ibid., viii.
21. Ibid., 11.
22. Ibid., 13.
23. Gilbert Meilaender, *Friendship: A Study in Theological Ethics* (Notre Dame, IN: University of Notre Dame Press, 1981) as quoted in Schwehn, *Exiles from Eden*, 19.
24. For some cautions from cognitive psychology, see David G. Myers, *Intuition: Its Powers and Perils* (New Haven, CT: Yale University Press, 2002), and Timothy D. Wilson, *Strangers to Ourselves: Discovering the Adaptive Unconscious* (Cambridge, MA: Belknap Press of Harvard University Press, 2002).
25. Michael Andre Bernstein, *Foregone Conclusions: Against Apocalyptic History* (Berkeley: University of California Press, 1994).
26. Ibid., 3–4.
27. See, e.g., Lee Hardy, *The Fabric of this World* (Grand Rapids, MI: Eerdmans, 1990).

## 7 Inclinations

1. Ian Hacking, *The Social Construction of What?* (Cambridge, MA: Harvard University Press, 1999), and John R. Searle, *The Construction of Social Reality* (New York: Free Press, 1995), John R. Searle, *Mind, Language, and Society: Philosophy in the Real World* (New York: Basic Books, 1998).
2. "In the ordinary sense of the word, to *interpret* is to bring out the meaning of something that would not otherwise be clear. The object could be a world, a text, or an action. It could be a period of history or it could be a gesture." William E. Paden, *Interpreting the Sacred: Ways of Viewing Religion* (Boston: Beacon Press, 1992; emphasis in the original), 9.
3. Sometimes *interpretation* is connected with scientific programs of causal explanation; usually, however, it is associated with the humanistic explication of meanings. These are quite different tasks. Some would even reserve the term *explanation* for science and *interpretation* for humanistic topics. Ibid., 10.
4. A *descriptive theory* is a generalization "to the effect that some or all members of a set of entities possess certain properties or stand in certain relations." Descriptive theories may be predictive or explanatory, and the prediction or explanation may be nomothetic-deductive or statistical. A *normative theory* specifies what ought to be done. Foundationalism is a normative theory that specifies how theorizing should be done. Nicholas Wolterstorff, *Reason within the Bounds of Religion*, 2nd ed. (Grand Rapids, MI: Eerdmans, 1984), 63.
5. Random House Unabridged Dictionary.
6. Ronald F. Thiemann, *Religion in Public Life: A Dilemma for Democracy, A Twentieth Century Fund Book* (Washington, DC: Georgetown University Press, 1996), 134.

7. I speak of choice, but individual scholars may experience their selection as simply the solely "right" approach to the evidence available and the questions asked. Background beliefs, I submit, play a central role in producing this sense of "rightness."
8. For these broad categories, see the chapters "Religious Formation" and "Disciplinary Formation."
9. Or nondiscursive or "tacit." On this last term, see Michael Polanyi, in *Knowing and Being*, ed. Margorie Grene (Chicago: The University of Chicago Press, 1969), 123–207; Michael Polanyi, *Personal Knowledge: Towards a Post-Critical Philosophy* (Chicago: The University of Chicago Press, 1962); and Michael Polanyi and Harry Prosch, *Meaning* (Chicago: University of Chicago Press, 1975).
10. Wolterstorff, *Reason*.
11. Ibid., 67–68.
12. For this helpful distinction between "authentic" and "actual" Christian commitment, see ibid., 71–75.
13. Ibid., 95–96.
14. Ibid., 99.
15. See the discussion of *bricolage* in "Narrative Identity."
16. On the risks of self-deception in narrative accounts about one's self, see, e.g., Timothy D. Wilson, *Strangers to Ourselves: Discovering the Adaptive Unconscious* (Cambridge, MA: Belknap Press of Harvard University Press, 2002) and the literature cited therein.
17. For an extended discussion of the role of emotions in thought, and an overview of the objections to and some danger in the heuristic I'm suggesting, see, among others, William E. Connolly, *Why I Am Not a Secularist* (St. Paul, MN: University of Minnesota Press, 1999); David G. Myers, *Intuition: Its Powers and Perils* (New Haven, CT: Yale University Press, 2002); Martha C. Nussbaum, *Upheavals of Thought: The Intelligence of Emotions* (New York: Cambridge University Press, 2001); and Wilson, *Strangers to Ourselves*.
18. See, e.g., Nussbaum, *Upheavals of Thought*.
19. It should be noted that the standards extend well beyond the question of "fit" between "evidence" and "interpretation" or "explanation." Again with significant variation among disciplines, one also encounters standards regarding "simplicity," "coherence," "heuristic fertility or suggestiveness," "robustness," "aesthetics," "simplicity or elegance," "how interesting or stimulating an interpretation may be," "ability of being replicated," "falsifiability," "logical consistency," "meaningfulness," "predictive success," "fidelity to past theory," and so on and so forth. I have placed quotation marks around these broad standards to remind us that definitions for these standards (and their applicability or significant) vary both within a disciplinary community and from discipline to discipline. Further, not only can these standards vary among and within disciplinary communities, but they also may trade off against each other in complicated ways. Is, e.g., heuristic fertility more important than simplicity or elegance in developing theories in particle physics? Should aesthetic considerations count for more than strict logical consistency in literary analysis? What is the role of paradox? One of such interesting questions worth discussing is

how do you choose among such standards when the field itself does not make the choice for you?

20. See Robert H. Nelson, *Economics as Religion: From Samuelson to Chicago and Beyond* (University Park, PA: The Pennsylvania State University Press, 2001).
21. See Mary Midgley, *Science and Poetry* (London: Routledge, 2001).
22. See Paul Vitz, *Psychology as Religion: The Cult of Self-Worship*, 2nd ed. (Grand Rapids, MI: Eerdmans, 1994).
23. Midgley, *Science and Poetry*, 66.
24. Antonio R. Damasio, *Descartes' Error: Emotion, Reason, and the Human Brain* (New York: Avon, 1994).
25. See especially Polanyi, *Knowing and Being*; Polanyi, *Personal Knowledge*; Polanyi and Prosch, *Meaning*.
26. Thomas Nagel, *The View from Nowhere* (Oxford: Oxford University Press, 1986).

## 8 Community Warrant

1. For a rather extreme but forthright example of the type, see David Bromwich, *Politics by Other Means: Higher Education and Group Thinking* (New Haven, CT: Yale University Press, 1992).
2. Alasdair MacIntyre, *After Virtue: A Study in Moral Theory*, 2nd ed. (South Bend, IN: University of Notre Dame Press, 1984), 190.
3. See Richard J. Bernstein, *The New Constellation: The Ethical-Political Horizons of Modernity Postmodernity* (Cambridge, MA: The MIT Press, 1992), 48, 328, 36.
4. For a forthright declaration and defense of this position, see Robert B. Brandom, ed., *Rorty and His Critics* (Malden, MA: Blackwell Publishers, 2000); Richard Rorty, *Philosophy and Social Hope* (New York: Penguin, 1999); Richard Rorty, *Philosophy and the Mirror of Nature* (Princeton, NJ: Princeton University Press, 1979); and Richard Rorty, "Religion as Conversation-Stopper (1994)," in *Philosophy and Social Hope* (London: Penguin Books, 1999), 168–74.
5. Fidelity can be the deciding standard for scholars working within a religious tradition at a church-related institution of higher education. See the brief discussion in the chapter "Academic Freedom" regarding Catholic and Reformed understanding of academic freedom.
6. John R. Searle, *The Construction of Social Reality* (New York: Free Press, 1995), 12–13.
7. Recall MacIntyre's aphorism, "A living tradition then is an historically extended, socially embodied argument, and an argument precisely in part about the goods which constitute that tradition." MacIntyre, *After Virtue*, 222.
8. My sources for this position are, above all, Richard Bernstein, Alasdair MacIntyre, Jeffrey Stout, Ronald Thiemann, and several others. Naturally, these different scholars don't agree with each other on each and every point, but they do agree to a large extent on the key historicist, nonfoundational, and fallibilist convictions that underlie this way of understanding truth and justification. Richard J. Bernstein, *Beyond Objectivism and Relativism: Science, Hermeneutics, and Praxis* (Philadelphia: University of Pennsylvania Press, 1983); Bernstein,

*The New Constellation*; Richard J. Bernstein, "Pragmatism, Pluralism, and the Healing of Wounds (1988)," in *Pragmatism: A Reader*, ed. Louis Menand (New York: Vintage, 1997), 382–401; Richard J. Bernstein, "Religious Concerns in Scholarship: Engaged Fallibilism in Practice," in *Religion, Scholarship, & Higher Education: Perspectives, Models, and Future Prospects*, ed. Andrea Sterk (Notre Dame, IN: University of Notre Dame Press, 2002), 150–58; MacIntyre, *After Virtue*; Alasdair MacIntyre, *Three Rival Versions of Moral Enquiry: Encyclopeaedia, Genealogy, and Tradition* (South Bend, IN: University of Notre Dame Press, 1990); Alasdair MacIntyre, *Whose Justice? Which Rationality?* (Notre Dame, IN: University of Notre Dame Press, 1988); Nancey Murphy, *Anglo-American Postmodernity: Philosophical Perspectives on Science, Religion, and Ethics* (Boulder, CO: Westview Press, 1997); Nancey Murphy, *Beyond Liberalism and Fundamentalism: How Modern and Postmodern Philosophy Set the Theological Agenda* (Valley Forge, PA: Trinity Press International, 1996); Jeffrey Stout, *Ethics After Babel: The Languages of Morals and Their Discontents* (Boston: Beacon Press, 1988); Ronald F. Thiemann, *Constructing a Public Theology* (Louisville, KY: Westminster/John Knox Press, 1991); and Ronald F. Thiemann, *Religion in Public Life: A Dilemma for Democracy, A Twentieth Century Fund Book* (Washington, DC: Georgetown University Press, 1996).

9. For example, see John T. Noonan, Jr., *A Church That Can and Cannot Change* (Notre Dame, IN: University of Notre Dame Press, 2005).
10. Thiemann, *Religion in Public Life*, 154.
11. Ibid., 155.
12. Ibid., 155–56.
13. Speaking about the admissibility of Christian perspectives in the public sphere, Thiemann makes a similar point. Ibid., 156.
14. Bernstein, *Beyond Objectivism and Relativism*; Bernstein, "Pragmatism," and Bernstein, "Religious Concerns."

## 9 Academic Freedom

1. I owe this insight and its development to an article by historian Thomas Haskell. Thomas L. Haskell, "Justifying the Rights of Academic Freedom in the Era of Power/Knowledge," in *The Future of Academic Freedom*, ed. Louis Menand (Chicago: The University of Chicago Press, 1996), 43–90. For the historical introduction in "Disciplinary Formation" that I pick up here, I rely principally on Jon H. Roberts and James Turner, *The Sacred and the Secular University* (Princeton, NJ: Princeton University Press, 2001), and Julie A. Reuben, *The Making of the Modern University: Intellectual Transformation and the Marginalization of Morality* (Chicago: The University of Chicago Press, 1996).
2. The 1915 *Declaration of Principles* may be found in AAUP, "The 1915 Declaration of Principles," *Academic Freedom and Tenure: A Handbook of the American Association of University Professors*, ed. Louis Joughin (Madison: University of Wisconsin Press, 1969), 155–76.

3. Ibid., 161–62.
4. Ibid., 162.
5. Ibid.
6. Ibid.
7. Ibid., 163.
8. The following is taken from the *1940 Statement of Principles on Academic Freedom and Tenure With 1970 Interpretive Comments* found at http://www.aaup.org/statements/Redbook/1940stat.htm. It is also reproduced in Joughin, ed., *Academic Freedom and Tenure*, 373–83.
9. From the vast literature on academic freedom in America, I have drawn principally on Anthony J. Diekema, *Academic Freedom & Christian Scholarship* (Grand Rapids, MI: Eerdmans, 2000); Neil Hamilton, *Zealotry and Academic Freedom: A Legal and Historical Perspective* (New Brunswick, NJ: Transaction Publishers, 1995); Louis Menand, ed., *The Future of Academic Freedom* (Chicago: The University of Chicago Press, 1996); and Robert K. Poch, *Academic Freedom in American Higher Education: Rights, Responsibilities and Limitations*, ed. Jonathan D. Fife, *ASHE-ERIC Higher Education Reports—Report 4* (Washington, DC: The George Washington University, School of Education and Human Development, 1993). The classic history is Richard Hofstadter and Walter P. Metzger, *The Development of Academic Freedom in the United States* (New York: Columbia University Press, 1955).
10. Poch, *Academic Freedom*, 12.
11. Or liberty; in this essay in political philosophy, Berlin employs the two terms interchangeably. See Isaiah Berlin, "Two Concepts of Liberty (1958)," in *The Proper Study of Mankind: An Anthology of Essays/Isaiah Berlin*, ed. Henry Hardy and Roger Hausheer (New York: Farrar, Straus and Giroux, 1997), 194. I also draw from Berlin the insight that goods may clash and be in some ways incompatible with each other. On this point, with respect to academic freedom, see also Louis Menand, "The Limits of Academic Freedom," in *The Future of Academic Freedom*, ed. Louis Menand (Chicago: The University of Chicago Press, 1996), 14, and Joan W. Scott, "Academic Freedom as an Ethical Practice," in *The Future of Academic Freedom*, ed. Louis Menand (Chicago: The University of Chicago Press, 1996), 163–80.
12. This freedom from outside interference may be limited to matters pertinent to one's discipline.
13. Berlin, "Two Concepts of Liberty (1958)," 203.
14. Ibid., 204.
15. Ibid., 241.
16. Ibid., 239. This is a key aspect of "objective pluralism" a doctrine developed by Berlin. Robert Audi, ed., *The Cambridge Dictionary of Philosophy*, 2nd ed. (Cambridge, UK: Cambridge University Press, 1999), 85–86.
17. Isaiah Berlin, "The Pursuit of the Ideal (1988)," in *The Proper Study of Mankind: An Anthology of Essays/Isaiah Berlin*, ed. Henry Hardy and Roger Hausheer (New York: Farrar, Straus and Giroux, 1997), 11.
18. Ibid., 10–11.
19. AAUP, "1915 Declaration," reprinted in Hamilton, *Zealotry*, 357–71.

20. AAUP, "1915 Declaration," reprinted in Hamilton, *Zealotry*, 366.
21. Ibid.
22. Ibid., 367.
23. Ibid.
24. AAUP, *1940 Statement of Principles on Academic Freedom and Tenure with 1970 Interpretive Comments* (1940, 1970 [cited June 27, 2005]); available from http://www.aaup.org/statements/Redbook/1940stat.htm.
25. The United States National Student Association (now the United States Student Association), the Association of American Colleges (now the Association of American Colleges and Universities), the National Association of Student Personnel Administrators, and the National Association of Women Deans and Counselors (now the National Association for Women in Education).
26. AAUP, *Joint Statement on Rights and Freedoms of Students* (1967 [cited June 27, 2005]); available from http://www.aaup.org/statements/Redbook/Studentrights.htm.
27. Hamilton, *Zealotry*, 366.
28. From the "About Us" page at http://www.studentsforacademicfreedom.org.
29. Students for Academic Freedom (SAF), *Students for Academic Freedom Mission and Strategy* (2004 [cited June 28, 2005]); available from http://www.studentsforacademicfreedom.org/essays/pamphlet.html.
30. For a list of the national and state legislation texts with links, see http://www.studentsforacademicfreedom.org/reports/NationalandStateLegislation.htm
31. American Legislative Exchange Council, *Model Legislation: [House/Senate] Joint Resolution Academic Bill of Rights for Higher Education* (2004 [cited June 28, 2005]); available from http://www.studentsforacademicfreedom.org/archive/May2004/Academic%20Bill%20of%20Rights%20Reso..pdf.
32. For example, "academic freedom and intellectual diversity are values indispensable to an American university," Ibid. See also Students for Academic Freedom, *Academic Bill of Rights* (2004 [cited June 28, 2005]); available from http://www.studentsforacademicfreedom.org/abor.html, and Students for Academic Freedom, *The Student Bill of Rights* (2004 [cited June 28, 2005]); available from http://www.studentsforacademicfreedom.org/essays/sbor.html.
33. SAF, *Mission and Strategy*.
34. SAF, *Academic Bill of Rights*.
35. Hamilton, *Zealotry*, 366.
36. Education Committee, *Senate Staff Analysis and Economic Impact Statement for Bill Sb 2126* (2005 [cited June 28, 2005]); available from http://www.flsenate.gov/data/session/2005/Senate/bills/analysis/pdf/2005s2126.ed.pdf. Partly contested by house staff analysis, which conceded that the bill "could increase the role of administrators or the courts in determining whether or not a student's or faculty member's freedom has been infringed" but denied that the bill created "a statutory cause of action for students, instructors, or faculty who feel their rights have been infringed." Education Council, *House of Representatives Staff Analysis (Bill: Hb 837)* (2005 [cited June 28, 2005]); available from http://www.flsenate.gov/data/session/2005/House/bills/ analysis/pdf/ h0837c.EDC.pdf.

37. James Vanlandingham, *Capitol Bill Aims to Control "Leftist" Profs: The Law Could Let Students Sue for Untolerated Beliefs* (2005 [cited June 28, 2005]); available from http://www.alligator.org/pt2/050323freedom.php.
38. *Conservative "Academic Bill of Rights" Limits "Controversial Matter" in Classroom (April 6, 2005)* (2005 [cited June 28, 2005]); available from http://www.democracynow.org/print.pl?sid = 05/04/06/1421208.
39. The Academic Bill of Rights, and the proposed Florida legislation, seems to single out the humanities, social science, and the arts for special attention, stating that "the fostering of a plurality of serious scholarly methodologies and perspectives should be a significant institutional purpose" in these academic areas. The natural sciences are not mentioned, but presumably still fall under the rights assigned to students, including the "right to expect a learning environment in which they will have access to a broad range of serious scholarly opinion pertaining to the subjects they study." Robert Baxley, *A Bill to Be Entitled an Act Relating to Student and Faculty Academic Freedom* (2005 [cited June 28, 2005]); available from http://www.flsenate.gov/data/session/2005/House/bills/billtext/pdf/h083700.pdf. Cf, principle 5 in the Academic Bill of Rights Freedom, *Academic Bill of Rights*.
40. See the insightful analysis in Stanley Fish, " 'Intellectual Diversity': The Trojan Horse of a Dark Design," *The Chronicle of Higher Education*, February 13, 2004, B13.
41. On June 23, 2005, the American Council on Education (ACE) issued a *Statement on Academic Rights and Responsibilities*, which, while not mentioning the *Academic Bill of Rights* by name, is clearly intended as a response. This *Statement* is endorsed by 27 other academic associations including the American Association of University Professors. The *Statement* stresses that academic freedom and intellectual pluralism are not only "central principles of American higher education" but also "complex topics with multiple dimensions that affect both students and faculty." Moreover, the diversity of American colleges and universities make it impossible to create a single standard regarding these issues. "Individual campuses," the *Statement* explains, "must give meaning and definition to these concepts within the context of disciplinary standards and institutional mission." Even so, the *Statement* concedes that there are some overarching principles that deserve to be affirmed. It starts with the great diversity of institutions that makes up American higher education. This diversity needs to be valued and protected. It continues with the assertion that colleges and universities "should welcome intellectual pluralism and the free exchange of ideas" that should occur in an "environment characterized by openness, tolerance and civility." But when it comes to treat what the Academic Bill of Rights called "intellectual diversity" and the implication who decides what's uncertain or unsettled in a discipline and who decides on the merits of ideas, theory, arguments, or views to be shared with students, the ACE *Statement* comes out four-square for institutional and professional autonomy:

    > The validity of academic ideas, theories, arguments and views should be measured against the intellectual standards of relevant academic and professional disciplines. Application of these intellectual standards does not mean

> that all ideas have equal merit. The responsibility to judge the merits of competing academic ideas rests with colleges and universities and is determined by reference to the standards of the academic profession as established by the community of scholars at each institution.

And the *Statement* closes with the assertion that "government's recognition and respect for the independence of colleges and universities is essential for academic and intellectual excellence." In turn, colleges and universities have "a particular obligation to ensure that academic freedom is protected for all members of the campus community and that academic decisions are based on intellectual standards consistent with the mission of each institution." American Council on Education, *Statement on Academic Rights and Responsibilities* (June 23, 2005 2005 [cited June 28, 2005]); available from http://www.acenet.edu/AM/Template.cfm?Section=Search&template=/CM/ContentDisplay.cfm&ContentID=10672.

While David Horowitz called this *Statement* a "major victory" for his campaign, on the central issue of "intellectual diversity" in curricula, reading lists, and presentation of "dissenting viewpoints," it is hard to see how this furthers his agenda, for the ACE *Statement* returns the final decision to the disciplinary communities of practice and their local institutional homes, where they have been lodged at least in principle since 1915. Scott Jaschik, *Detente with David Horowitz* (June 23, 2005 [cited June 28, 2005]); available from http://www.insidehighered.com/news/2005/06/23/statement.

42. Hamilton, *Zealotry*, 358.
43. Poch, *Academic Freedom*, 59.
44. For a thoughtful treatment of these issues from a Reformed Calvinist perspective, see Diekema, *Academic Freedom & Christian Scholarship*.
45. For an extended discussion on this and related issues, see Poch, *Academic Freedom*, 58–65, and Diekema, *Academic Freedom & Christian Scholarship*.
46. Quoted in Poch, *Academic Freedom*, 60.
47. This brief summary cannot do justice to the complications and continuing debates elicited by *Ex corde ecclesiae* (1990). For a succinct but sophisticated overview of the issues, see James L. Heft, SM, "Academic Freedom," *New Catholic Encyclopedia* (2003). I draw the quote from this article, pp. 54–55. See also Kenneth W. Kemp, *What Is Academic Freedom*? (November 15, 2000 [cited July 5, 2005]); available from http://courseweb.stthomas.edu/kwkemp/Papers/AF.pdf.
48. For nuance and justification, see Diekema, *Academic Freedom & Christian Scholarship*, 44–81.

## 10 RETICENCE

1. For some scholars, deep political convictions function in their thinking much as religious or spiritual convictions do in other scholars.
2. For example, George Marsden chides the Yale historian Skip Stout and the Princeton sociologist Robert Wuthnow (among others) for interpretations that

they advance that reflect their evangelical Christian commitment without explicitly acknowledging that commitment. See George M. Marsden, *The Outrageous Idea of Christian Scholarship* (New York: Oxford University Press, 1997), 65–68, 70–74.

3. For example, Marsden adds "naturalistic reductionism" to this list; ibid., 72–77.
4. The nub of the disagreement rests on the "moral status" of the embryo. See, e.g., The President's Council on Bioethics, *Monitoring Stem Cell Research* (Washington, DC: United States Government, 2004), 14.
5. Or they may be assumptions that are justified instrumentally. That is, no claim is made that the assumption depicts reality. Rather the claim made is that the model, whatever the validity of this particular assumption about human beings, works well in predicting human activity. For a critical review of Rational Choice theory in political science, see Jeffrey Friedman, ed., *The Rational Choice Controversy: Economic Models of Politics Reconsidered* (New Haven, CT: Yale University Press, 1996), and Donald P. Green and Ian Shapiro, *Pathologies of Rational Choice Theory: A Critique of Applications in Political Science* (New Haven, CT: Yale University Press, 1994).
6. Or "a coherent and comprehensive picture of the whole world," William Seager, "Metaphysics, Role in Science," in *A Companion to the Philosophy of Science*, ed. W. H. Newton-Smith (Malden, MA: Blackwell Publishers, 2000), 290.
7. *The Encyclopedia of Philosophy* (New York: Macmillan, 1967) 5: 448. For more recent attempts at short characterization, and a guide to some of the major literature, see Ronald N. Giere, "Naturalism," in *A Companion to the Philosophy of Science*, ed. W. H. Newton-Smith (Malden, MA: Blackwell Publishers, 2000), 308–10; Alan Lacey, "Naturalism," in *The Oxford Companion to Philosophy*, ed. Ted Honderich (New York: Oxford University Press, 1995), 604–06; and Philip Pettit, "Naturalism," in *A Companion to Epistemology*, ed. Jonathan Dancy and Ernest Sosa (Malden, MA: Blackwell Publishers, 1992), 296–97.
8. Stout makes this suggestion in analyzing the interview techniques of *Habits of the Heart* [Robert N. Bellah, Richard Madsen, William M. Sullivan, Ann Swidler, and Steven M. Tipton, *Habits of the Heart: Individualism and Commitment in American Life* (Berkeley: University of California Press, 1985).]. Jeffrey Stout, *Ethics After Babel: The Languages of Morals and Their Discontents* (Boston: Beacon Press, 1988), 195–96.
9. Of course, even in such a case the person who claims that lying is wrong may still be open to discussing whether lying is wrong when, say, engaged in by prisoners of war who are trying to deceive the enemy. But now he's discussing contextual exceptions to the general moral maxim.
10. When religious apologists argue that religious belief is inherently nonrational, perhaps in a misguided belief that such a claim will protect the faith from corrosive rational scrutiny, they are (from my perspective, unwisely and unnecessarily) ceding ground to those who wish to bar religious or spiritual claims from academic discourse. As theologian Ronald Thiemann argues regarding religion's proper role in public political discourse, religious or spiritual commitments are as open to external scrutiny or critique as any other fundamental commitments.

See Ronald F. Thiemann, *Religion in Public Life: A Dilemma for Democracy, A Twentieth Century Fund Book* (Washington, DC: Georgetown University Press, 1996), 154–59.

11. This argument is commonly encountered in discussions regarding the suitability of religious claims in public political discourse. For examples and bibliography, see Jeffrey Stout, *Democracy and Tradition* (Princeton, NJ: Princeton University Press, 2004), and Thiemann, *Religion in Public Life.*
12. See, e.g., Friedman, ed., *The Rational Choice Controversy*; Green and Shapiro, *Pathologies of Rational Choice Theory*; and Daniel Kahneman and Amos Tversky, eds., *Choices, Values, and Frames* (New York: Cambridge University Press, 2000).
13. This may even hold for moral philosophers and ethicists. See Moody-Adams's (challenging) claim, Michele M. Moody-Adams, *Fieldwork in Familiar Places: Morality, Culture, and Philosophy* (Cambridge, MA: Harvard University Press, 1997), 169ff.
14. For an illustration of this point, see the overview of the various arguments for and against embryonic stem cell research. President's Council on Bioethics, *Monitoring Stem Cell Research.*
15. The literature on this topic is vast; for a brief overview, see Ernan McMullin, "Values in Science," in *A Companion to the Philosophy of Science*, ed. W. H. Newton-Smith (Malden, MA: Blackwell Publishers, 2000), 550–60. For a trenchant critique, see Hilary Putnam, *The Collapse of the Fact/Value Dichotomy and Other Essays* (Cambridge, MA: Harvard University Press, 2002). For the fact-value, is-ought distinction within ethics and the continuing dispute, see the overview in Louis P. Pojman, *Ethics: Discovering Right and Wrong* (Belmont, CA: Wadsworth Publishing Company, 1995), 186–219.
16. For a thoughtful exploration of such issues, see Moody-Adams, *Fieldwork in Familiar Places.*
17. See, e.g., John Rawls, *Political Liberalism*, vol. 4, *The John Dewey Essays in Philosophy* (New York: Columbia University Press, 1996); Stout, *Democracy and Tradition*; and Thiemann, *Religion in Public Life*, and their bibliographies.
18. Whatever the theoretical merits of a ban in the public arena or in the academy against the deployment of religious or spiritual claims, one might also ask whether it is practicable. Certainly, the theorists' ban on religious claims in the public arena has had little, if any, effect on actual political debates within national, state, or even local politics. To be sure, the academy may be less unruly in this regard, and a ban backed by collegial peer pressure has been reasonably effective at secular institutions of higher education. See "Cautionary Tales" for an overview of why this ban has made a certain sense.
19. For an application of this point to disagreements on morality, see Moody-Adams, *Fieldwork in Familiar Places*, 110–12.
20. George A. Lindbeck, *The Nature of Doctrine: Religion and Theology in a Postliberal Age* (Philadelphia: The Westminster Press, 1984), 32.
21. It makes no difference to the argument that I'm advancing here, but it may be worth considering whether nonreligious comprehensive interpretive schemes may have their own, secular myths, narratives, and rituals. See the discussion in "Cautionary Tales."

22. Various "postmodern" comprehensive schemes in the humanities will challenge naturalism and, for that matter, the notion of comprehensive schemes. I chose to focus here on naturalism to illustrate the larger point I am trying to make. For two sympathetic readings of some postmodern alternatives, see Richard J. Bernstein, *The New Constellation: The Ethical-Political Horizons of Modernity Postmodernity* (Cambridge, MA: The MIT Press, 1992), and Barbara Herrnstein Smith, *Belief & Resistance: Dynamics of Contemporary Intellectual Controversy* (Cambridge, MA: Harvard University Press, 1997).
23. For three recent attempts at brief characterization, see Giere, "Naturalism," Lacey, "Naturalism," and Pettit, "Naturalism."
24. Simon Blackburn, ed., *The Oxford Dictionary of Philosophy* (New York: Oxford University Press, 1994), 255.
25. Lacey, "Naturalism," 604.
26. There are other major candidates, e.g., abstract entities, universals, and, of course, spirits, "world-spirit," and God or gods.
27. For an extended and nuanced example and justification for this approach, see Alasdair MacIntyre, *Whose Justice? Which Rationality?* (Notre Dame, IN: University of Notre Dame Press, 1988).
28. In formerly communist countries, you may find similar suspicions regarding Marxism. And various capitalist ideologies have been known in the past to threaten academic freedom; in fact, today's academic freedom may have arisen more in response to aggressive laissez-faire capitalism than to aggressive Christianity. For a variety of views on religion (and other "outside" forces) and academic freedom, see Anthony J. Diekema, *Academic Freedom & Christian Scholarship* (Grand Rapids, MI: Eerdmans, 2000); Thomas L. Haskell, "Justifying the Rights of Academic Freedom in the Era of Power/Knowledge," in *The Future of Academic Freedom*, ed. Louis Menand (Chicago: The University of Chicago Press, 1996), 43–90; Richard Hofstadter and Walter P. Metzger, *The Development of Academic Freedom in the United States* (New York: Columbia University Press, 1955); and Louis Menand, "The Limits of Academic Freedom," in *The Future of Academic Freedom*, ed. Louis Menand (Chicago: The University of Chicago Press, 1996), 3–20.
29. See, e.g., Marsden, *Outrageous Idea*, 3743, and Warren A. Nord, *Religion & American Education: Rethinking a National Dilemma* (Chapel Hill: The University of North Carolina Press, 1995), 241–47. For the political realm, see, among others, Stephen L. Carter, *The Culture of Disbelief: How American Law and Politics Trivialize Religious Devotion* (New York: Basic Books, 1993); Stout, *Democracy and Tradition*; and Thiemann, *Religion in Public Life*.
30. For just the opposite approach, see Merold Westphal, *Suspicion & Faith: The Religious Uses of Modern Atheism* (Grand Rapids, MI: Eerdmans, 1993).
31. Carter, *The Culture of Disbelief.*
32. Richard Rorty, "Religion as Conversation-Stopper (1994)," in *Philosophy and Social Hope* (London: Penguin Books, 1999), 168–69.
33. John R. Searle, *Mind, Language, and Society: Philosophy in the Real World* (New York: Basic Books, 1998), 34.

34. Peter L. Berger, ed., *The Desecularization of the World: Resurgent Religion and World Politics* (Grand Rapids, MI: Eerdmans, 1999), 2.
35. Ibid., 11–12.

## 11 In the Classroom

1. Warren A. Nord, *Religion & American Education: Rethinking a National Dilemma* (Chapel Hill: The University of North Carolina Press, 1995).
2. Donald P. Green and Ian Shapiro, *Pathologies of Rational Choice Theory: A Critique of Applications in Political Science* (New Haven, CT: Yale University Press, 1994), xi.
3. For a general, but not explicitly religious, critique, see ibid.
4. See Alexander Astin and Helen Astin, "The Spiritual Life of College Students: A National Study of College Students' Search for Meaning and Purpose," in *Spirituality in Higher Education* (Los Angeles: UCLA Higher Education Research Institute, 2005), and Christian Smith and Melinda Lundquist Denton, *Soul Searching: The Religious and Spiritual Lives of American Teenagers* (New York: Oxford University Press, 2005).
5. Nord, *Religion & American Education*, 210.
6. Ibid., 211.
7. There is a logic to natural inclusion in introductory courses, but a case can also be made that upper-division courses would allow faculty and students to explore in more satisfying depth the mutual bearing of disciplinary issue and religion.
8. Robert H. Nelson, *Economics as Religion: From Samuelson to Chicago and Beyond* (University Park, PA: The Pennsylvania State University Press, 2001), 69–70.
9. Samuel P. Huntington, *The Clash of Civilizations and the Remaking of World Order* (New York: Simon & Schuster, 1996).
10. Benjamin Barber, *Jihad Vs. McWorld* (New York: Times Books, 1995).
11. Nord himself would generally accept these constraints on the broader concept (personal conversation).
12. See, e.g., the fine piece by the Brown University molecular biologist Kenneth Miller. Kenneth R. Miller, *Finding Darwin's God: A Scientist's Search for Common Ground between God and Evolution* (New York: Cliff Street Books, 1999).
13. The literature on this topic is vast. For its history, see Ronald L. Numbers, *The Creationists: The Evolution of Scientific Creationism* (New York: Knopf, 1992). Some recent or important contributions include Barbara Forrest and Paul R. Gross, *Creationism's Trojan Horse: The Wedge of Intelligent Design* (New York: Oxford University Press, 2004); Michael Ruse, *The Evolution-Creation Struggle* (Cambridge, MA: Harvard University Press, 2005); and Eugenie C. Scott, *Evolution Vs. Creationism: An Introduction* (Westport, CT: Greenwood Press, 2004).
14. For the lay of the land on this topic, see Jeffrey H. Schwartz, *Sudden Origins: Fossils, Genes, and the Emergence of Species* (New York: John Wiley, 1999).

15. See, e.g., Susan Oyama, *The Ontogeny of Information: Developmental Systems and Evolution*, 2nd rev. and exp. ed. (Durham, NC: Duke University Press, 2000).
16. See, e.g., Roger Lundin, ed., *Disciplining Hermeneutics: Interpretation in Christian Perspective* (Grand Rapids, MI: Eerdmans, 1997).
17. W. James Bradley and Kurt C. Schaefer, *The Uses and Misuses of Data and Models: The Mathematization of the Human Sciences* (Thousand Oaks, CA: Sage Publications, 1998).
18. For a (variously persuasive) justification of this position, see Anthony J. Diekema, *Academic Freedom & Christian Scholarship* (Grand Rapids, MI: Eerdmans, 2000). See also the discussion in the chapter "Academic Freedom."
19. Diana L. Eck, *A New Religious America: How A "Christian Country" Has Now Become the World's Most Religiously Diverse Nation* (San Francisco: HarperSanFrancisco, 2001).
20. See, e.g., the accounts in Julie A. Reuben, *The Making of the Modern University: Intellectual Transformation and the Marginalization of Morality* (Chicago: The University of Chicago Press, 1996), and Jon H. Roberts and James Turner, *The Sacred and the Secular University* (Princeton, NJ: Princeton University Press, 2001).
21. See, e.g., Russell T. McCutcheon, *Critics Not Caretakers: Redescribing the Public Study of Religion* (Albany, NY: SUNY Press, 2001).
22. This expectation is discussed at greater length in the chapters "Disciplinary Formation" and "Academic Freedom."
23. Warren A. Nord and Charles C. Haynes, *Taking Religion Seriously across the Curriculum* (Nashville, TN: First Amendment Center, 1998).

## Conclusion

1. Alasdair MacIntyre, *After Virtue: A Study in Moral Theory*, 2nd ed. (South Bend, IN: University of Notre Dame Press, 1984), 222.

## Appendix 1 Advice for Seminar Leaders

1. For an eminently helpful discussion of what goes into a "healthy" conversation, see Douglas Stone, Bruce Patton, and Sheila Heen, *Difficult Conversations: How to Discuss What Matters Most* (New York: Viking, 1999).

# Index